Also by Norman Friedman

Bridging Science and Spirit:
Common Elements in David Bohm's Physics, the Perennial Philosophy and Seth

"I think *Bridging Science and Spirit* is one of the most insightful, comprehensive, and brilliant expressions of knowledge. I shall certainly use it as a reference guide. Many abstract ideas that I was not comfortable with are now more meaningful."

Deepak Chopra, M.D. - Author of *Ageless Body, Timeless Mind; The Seven Spiritual Laws of Success; The Way of the Wizard*

"*Bridging Science and Spirit* accomplishes a formidable task. This book will be a valuable research document for many years to come for those concerned with a perspective that honors both science and spirituality."

Fred Alan Wolf, Ph.D. - Author of *Taking the Quantum Leap; Parallel Universes; The Physics of the Soul*

"An insightful synthesis of the outer frontiers of theoretical physics with the deep revelations of mystical insight, throwing important light on the reality we live in."

Peter Russell - Author of *The Global Brain; The White Hole in Time; The Global Brain Awakens*

"A courageous and visionary work, consistent not only with good science but with humankind's perennial spiritual vision. This is a very important book."

Larry Dossey, M.D. - Author of *Recovering the Soul; Healing Words; Prayer is Good Medicine*

"*Bridging Science and Spirit* correctly asks, 'How does matter originate from consciousness?' This is the fundamental question of a growing body of literature regarding the new paradigm of an idealist, consciousness-based science. Norman Friedman has made an important and thoughtful contribution to this new science."

Amit Goswami, Ph.D. - Professor of Physics, University of Oregon. Author of *The Self-Aware Universe*

"Some future historian will, I feel confident, identify the 'Great Debate' of the twentieth century around the question: 'What is science going to do about consciousness?' I would surmise that Norman Friedman's book *Bridging Science and Spirit* will turn out to be a benchmark in that inquiry.... No one can read it without gaining some clarity on their own nature."

Willis Harman, Ph.D. - President, Institute of Noetic Sciences. Author of *Higher Creativity; An Incomplete Guide to the Future*

"...a revisioning of physics...a challenge to reductionistic models of nature...radical, yet stimulating and inspiring."

Stanley Krippner, Ph.D. - Distinguished Professor of Psychology, California Institute of Integral Studies

"Few have understood both science and metaphysics well enough to remove our blinders to their underlying similarities — and universal truths. Kudos to Norman Friedman for proving to be a master bridgebuilder."

Lynda Dahl - President, Seth Network International. Author of *Beyond the Winning Streak; Ten Thousand Whispers; The Wizards of Consciousness*

"...a masterful job of bringing innate psychic understanding and scientific knowledge together, brilliantly showing us how all is one and one is all."

Robert F. Butts - Husband of the late Jane Roberts, channel for **Seth**

"...a brilliant synthesis of ageless wisdom and modern science. Friedman's writing provides pictures for our minds so we can 'see' reality...."

Jacquelyn Small, MSSW - President, Eupsychia Inc. Author of *Awakening in Time; Transformers, the Artists of Self-Creation*

"Norman Friedman's special gift is his ability to bring out the imaginative significance of the new paradigm. Recommended for all who enjoy having their minds expanded."

Michael Grosso - Author of *Frontiers of the Soul; Soulmaker*

"I responded to this book with both admiration and appreciation for the task Norman Friedman set out to do and accomplished so successfully.... He has paved the way toward a deeper understanding of the role of consciousness in the creation and structure of matter."

Montague Ullman, M.D. - Clinical Professor of Psychiatry, Emeritus, Albert Einstein College of Medicine, NY.

THE HIDDEN DOMAIN

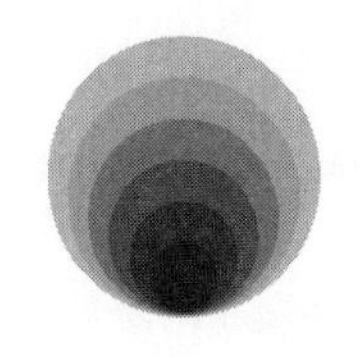

Home of the Quantum Wave Function, Nature's Creative Source

NORMAN FRIEDMAN

Published by

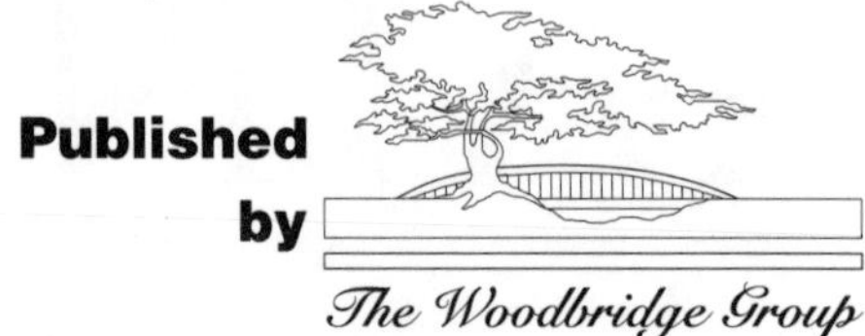

P.O. Box 849
Eugene, OR 97440
541/683-6731 / fax 541/683-1084
E-mail: woodbridge@jb.com

Cover design and illustration by
Lightbourne Images copyright © 1997

Library of Congress Cataloging-in-Publication Data

Friedman, Norman, 1926–
Hidden domain : home of the wave function, nature's creative source / by Norman Friedman
p. cm.
Includes bibliographical references and index.
ISBN 1-889964-09-3 (pbk.)
1. Wave functions. 2. Quantum theory. I. Title.
QC174.26.W3F75 1997
530.12'4—dc21 96-37816
CIP

DEDICATION:

This book is for my grandchildren, Carolyn, Becky, Jessie, Rachel, and Nick, with loving hope that their generation may find in the hidden domain a source of wonder, meaning, and inspiration.

ACKNOWLEDGMENTS:

To my wife and partner, Leah Friedman, I express my gratitude for the countless hours she spent listening to me, discussing ideas with me, as well as reading and rereading the manuscript. My secretary Carolyn Fortel, who has been my assistant for many years, did an outstanding job of typing and making needed corrections on the many versions of this book. My good friend and editor, Sara Jenkins, endured with great patience my initial resistance to many of her valuable and pertinent suggestions regarding structure and other editorial matters. Her contribution to the end product cannot be overstated; I am very grateful to her. To Marisa Lovesong, my publicist, I extend many thanks for her generous assistance and dedication to this project. Physicist Jack Sarfatti read the manuscript carefully and made helpful technical suggestions. I very much appreciate his effort. If any errors remain, I accept full responsibility.

CONTENTS:

PREFACE:

In writing this book on the quantum wave function I have drawn from many sources, primarily physicists, mathematicians, philosophers, and Seth. Here I would like to say a few words about Seth, the discarnate entity channeled by Jane Roberts, the poet and novelist who produced more than fifteen volumes from the channeled Seth material. In researching my previous book *Bridging Science and Spirit*, I came across a comment by a physicist who noted some similarity between the description of a concept in quantum theory called Heisenberg's potentia, which is described by the wave function, and a domain described by Seth. After studying this material in depth, I realized that Seth's ideas pertaining to physics originated from an intelligence with both a knowledge of science and a clear message about the nature of reality and human experience. I have utilized this unusual source throughout — wherever the information seems helpful and appropriate.

□ ○ □

Understanding the wave function normally requires some background in advanced mathematics. Since the vast majority of readers do not have such a background, I shall attempt explanations using images, metaphors, and analogies; these are solidly grounded in science to keep distortions to a minimum. Thus, this book is free from mathematical formulas except for one section, which requires some understanding of high school algebra.

In spite of my efforts to explain complex concepts in simple terms, I am aware that those without a science background may find some portions of this book (particularly Part 2) quite challenging. These portions are necessary to provide a firm scientific basis for the modeling interpretations and philosophical speculations that follow. My suggestion to readers who are having trouble is to read lightly (or skip) sections that are too difficult and continue on to those which are easier to grasp. Chapter 13, especially the summary, should be helpful. If you get the gist of what is being said — even though you may not completely follow the reasoning — you may well be rewarded with some satisfying insights into the nature of reality, ideas that are developed in Parts 3 and 4. For those wishing greater depth, the footnotes provide additional information regarding some of the scientific material.

If we are to comprehend the implications of modern physics, it is advisable to understand the problems that arose in classical physics which necessitated the development of relativity and quantum theories. To this end, the first part of the book offers a historical background on the world of Newton, followed by an overview of relativity and quantum theories. The second part of the book presents a detailed discussion of the wave function and its philosophical meaning. Part 3, Models of Reality, examines concepts both from physicists who have made contributions in this area and from Seth.

In the final part of the book, I allow myself to range freely and widely in considering how the models of reality discussed relate to disciplines outside science. It is my view, however, that the only scientific tool we have for understanding the level underlying time and space is the wave function, and it is to a nonmathematical examination of this hidden domain that the book is devoted.

INTRODUCTION:

Quantum theory tells us that all matter — and any part thereof — is associated with an unusual wave, a wave designated by the mathematical formulation called the wave function. The wave function has unquestionable accuracy and usefulness; but, oddly enough, the wave it describes cannot be detected by any ordinary means — which makes it not only enigmatic, but literally not of this world. We are therefore faced with the notion that there exists an entire realm of reality that lies outside our experience of the material world yet is intimately connected to it, an idea that is well established though still only vaguely understood by physicists.

Our knowledge of the wave function forces us to leave behind the comfortable thought that our three-dimensional world is all there is and that it changes through time in comprehensible and predictable ways. Instead we must acknowledge that, considered in its most fundamental aspect, that of elementary particles, our world requires a second level consisting of *potential* states, described by the wave function, which are brought into three-dimensional reality by an act of observation — an idea many consider metaphysical. (At least, this is the traditional view of many physicists, consistent with the Copenhagen interpretation of quantum theory.) So, while classical theory attempted to banish the mystical and the mysterious from physics, quantum theory has reinstated such notions in the form of abstract mathematical formalisms — one being the wave function — which address a reality that often seems to contradict our normal experience.

The wave function describes the state of an electron (or

any larger system) in terms of the *possibilities* open to it. All these possibilities coexist with a certain *probability* of coming into being, of entering our "real" world. Unlike some scientists, who believe the wave function is merely a mathematical formulation used for computation, I take the position that it is an accurate description of that unseen level of probabilities, a domain which I consider to be the fundamental source from which our objective world unfolds.

If we accept this idea — that the underlying level must be taken at face value as an essential ingredient in any total concept of reality — then we are presented with some fascinating challenges, for we are confronted with a strange world consisting of infinite choices available to us from a timeless realm filled with a vast supply of unmeasurable energy. This book is about that realm of possibilities and probabilities — the unseen level of reality designated by the wave function.

Nick Herbert, a creative thinker and writer on the subject of quantum physics, sees the world as composed of one thing, something he calls "quantumstuff."

> The world is one substance. As satisfying as this discovery may be to philosophers, it is profoundly distressing to physicists as long as they do not understand the nature of that substance. For if quantumstuff is all there is and you don't understand quantumstuff, your ignorance is complete. (Herbert, *Quantum Reality*, p. 40)*

I shall attempt to dispel some of the ignorance regarding quantumstuff by examining and discussing some of its salient features.

☐ One remarkable characteristic of the quantum world is wholeness. The realm of the wave function is not made up of separate parts, but is totally interconnected, interpenetrated, and intermeshed. Assuming everything is made of one substance, then each of us is part of some infinite, invisible web; what happens

*Full bibliographic references are given at the end of the book.

in one part of the universe reverberates everywhere, and strong instantaneous influences at some deeper level take place between distant points.

- When quantum theory informs us that an "act of observation" is needed to manifest, or make "real," a particular state from all the possible states that exist simultaneously in the world of the wave function, this implies that consciousness plays a major role in creating our reality. Though we have not yet learned how to incorporate consciousness into our science, I am convinced that we must come to terms with its primary importance.
- One of the most intriguing aspects of the unseen realm of the wave function is that it contains an incomprehensible amount of energy, which we have no way of measuring. I shall consider what this energy might be and speculate on its nature as a vast reservoir of light, not unlike that described by many mystics.
- I shall also examine such familiar concepts as motion, space, time, and matter and show how they can be understood in a broader context when viewed through the prism of the wave function. From the perspective of classical Newtonian physics, we are accustomed to seeing our reality as consisting of matter which obeys certain laws within our three-dimensional world of space and time. A different view — and a more complete picture — is possible when we also understand matter while it is still in the domain of the wave function as *potential* — as elusive as its nature in that context seems to be.

To talk about reality while ignoring that unperceived "other" world is like an iceberg describing itself with no awareness of the ocean from which it originated. Thus, we shall examine in some detail (though nonmathematically) that underlying ocean from which our reality arises, the home of the wave function, the realm I call *The Hidden Domain*.

THE PHYSICS BACKGROUND

The classical view, largely defined by Sir Isaac Newton, depicted the universe as a vast machine, made up of complex organizations of the smallest units of matter, often compared to small billiard balls. Such elementary particles were viewed as separate and independent parts that interacted in the three dimensions of space strictly according to precise laws which were considered universal and eternal. These laws state that the motion of every material body is theoretically predictable, making for a completely deterministic universe. Classical scientists considered human beings to be governed by these same orderly laws. The mechanical universe was basically indifferent to human goals and aspirations, pursuing its appointed course without apparent purpose. This mechanistic view was very influential and infiltrated all disciplines, including biology and psychology.

The scientific revolution at the beginning of this century raised the first serious question about the extent to which Newtonian principles apply — and thus began to alter the way we conceive of reality. The revolution started with two enormously successful theories of Albert Einstein. The first, the special theory of relativity, replaced the three-dimensional space of classical physics with a concept that is intuitively difficult to understand: four-dimensional space-time, in which space and time have no reality independent of each other. The theory also predicts unusual behavior of matter at speeds close to the velocity of light, leading to further nonclassical conceptions. The most famous of these is $E=mc^2$, meaning that matter and energy are one and the same, just existing in different states. Einstein's second theory is no less revolutionary. Referred to as the general theory of relativity, and extremely important in cosmology, it uses curved space-time to brilliantly describe Newtonian gravity. In this theory, particles are seen not as material objects but as something like vortices in a continual field. (A more accurate term would be singularity, which means, roughly, an edge to space-time.) As radical as these concepts are in challenging long-held assumptions, relativity theory nevertheless did not negate the strict determinism of the classical view. That was left to quantum theory.

Quantum theory differs profoundly from all the theories that preceded it. Niels Bohr, one of the founders of quantum theory, said pointedly that if you are not shocked by the world of the quantum, then you just do not understand it. Einstein, among others, found the theory so disturbing in its implications that he never came to terms with it, even though he helped formulate some of its basic precepts.

One reason so much has been written about the philosophical implications of modern physics is that it dramatically changes our understanding of the world of material objects. Not only did relativity theory change our relationship to the classical concepts of space and time, but quantum theory demonstrated that elementary particles (constituents of matter) and photons (constituents of light) can behave as *either* waves *or* particles, depending on the method used in an experiment. This meant that the classical scheme, in which particles are very definitely viewed as objects, was no longer tenable. Rather, in quantum theory and in relativity theory, elementary particles are seen as something much less tangible. Werner Heisenberg, another of the founders of quantum physics, said, "The elementary particles . . . form a world of potentialities or possibilities rather than one of things or facts" (Davies and Gribbin, p. 27).

To fully understand this broader picture of reality we must first examine the classical approach which describes our everyday world, then discuss relativity theory and the way in which it questioned certain classical assumptions. Finally, we must consider quantum theory and how it leads to the underlying realm of the wave function.

1 The classical view

Newtonian laws

In the latter part of the seventeenth century, Newton introduced the notion that the universe is governed by natural dynamic forces which he codified into laws. He based his laws — which operate in the same way regardless of location — not on some underlying principle, but from a description of what is observed in nature. These classical laws tell *how* an object behaves when subjected to various conditions, but they do not tell *why* it does so.

There are three laws of motion in the classical system.[1] We shall be concerned primarily with the first law, which involves the concept of inertia, and which states that an object continues in a state of rest or constant velocity unless acted on by an external force.[2] In other words, Newton said that motion is natural and does not need to be explained; only when motion is changed must we take note, and changes are caused by an outside force.[3] In fact, the way we know a force is present is through a change in motion, which indicates an interaction of some sort. Thus Newton's first law contains two fundamental implications: 1) that matter is inert, passive, inanimate, and so cannot itself bring about a change in its state; and 2) that to effect such a change requires an outside physical agent or force.

---

[1]The classical view is defined in this book as the physics that was accepted prior to Einstein's special theory of relativity, published in 1905.

[2]Newton's second law states that the force acting on a material body is proportional to the rate of change of the momentum of the body. This law provides a definition of force based on the inertial property of the body. The third law states that when one object exerts a force on another, there is an equal and opposite force exerted on the first object by the second object.

[3]Causality is one of the basic concepts in classical physics. However, Newton said that a cause cannot be discovered for uniform motion or for absolute rest.

In the classical world, matter, acted upon by physical forces, was the province of science, and mind was the province of religion. This division followed Descartes' conception of the universe as separated into mind and matter. There was no crossing this great divide; for example, the nonphysical phenomenon of information could not, in this scheme, influence matter. However, when science moved into ascendancy and the power of the church began to decline, attempts were made to explain animate or living matter, which raised the question of consciousness. The solution that still holds in the materialist worldview is epiphenomenalism, in which mind is seen as a secondary phenomenon of inanimate material, or as a property of certain configurations of matter. Consciousness is thus explained by reducing it to the interaction of matter and physical force fields.

Newton recognized that many events we observe seem to violate the first law of motion. If we throw an object into the air, it does not continue to move in a straight line but curves back to earth. An apple is at rest on a tree, yet when its stem is broken, it does not remain at rest but falls to earth. In response to such observations Newton developed the universal law of gravitation to complement his three laws of motion and to place them on a sound mathematical footing.

Simply stated, the law of gravitation says that any two objects seem to exhibit an attractive force. The force between them is proportional to the product of their masses and inversely proportional to the square of the distance between them. Since the earth is by far the larger and more massive object, an apple falls to earth and not toward a second apple. Furthermore, the force that pulls on the apple is the same one that keeps the moon revolving around the earth and the earth revolving around the sun. In other words, gravity is active everywhere in the universe. With gravity, Newton put in place the final law describing the interaction of matter, and with it the mechanistic view of the world became firmly established. Newton's laws of motion and gravitation remained the foundation of physics for almost three hundred years, until the early twentieth century when Einstein's theory of relativity demanded some revision of them.

The Newtonian concept of the world was indeed impressive. With three relatively simple laws, it described most of the behavior of inanimate matter, and with the law of gravitation, the explanation seemed complete. According to this scheme, the world is made up of material particles which precisely follow the dynamic laws Newton developed. In short, if the positions, velocities, and masses of these particles are known, then their positions, velocities, and masses at any future time can, in principle, be computed. Thus was created the philosophical concept of determinism: that is, all states of matter, both past and future, are locked into place by mathematical equations. Such a concept leaves no room for mind, and free will seems completely nonexistent.

Action at a distance

Within Newton's brilliant mathematical edifice is a problem that claims our attention: action at a distance. The problem arose because of the basic Newtonian tenet that material bodies can be reduced to particles. Thus, matter is seen as discontinuous, and one particle acting on another without physical contact requires action at a distance. Action at a distance was not an issue with Newton's first law of motion, but with the law of gravitation the question arises: can there be an interaction between two bodies if the intervening space is a vacuum? This challenging idea set the stage for the collapse of the mechanistic view. By the mid-1800s, other action-at-a-distance problems arose; it became apparent that magnetism and electrically charged bodies also were involved. In addition, light seemed to travel from the sun to the earth through an intervening vacuum.

Newton himself wisely bowed out of the controversy on how gravitational force is transmitted, though he did express an opinion:

> It is unconceivable that inanimate brute matter should, without the mediation of something else which is not material, operate upon and affect other matter without mutual contact. That one body can act upon another at a distance, through a vacuum, without the mediation of anything else . . . is to me so great an absurdity, that I believe no man, who has in philosophical matters a competent faculty of thinking, can ever fall into it.
> (Rucker, p. 72)

But for those who followed Newton, the problem had to be faced. From the materialist viewpoint, there were only two possibilities: either these forces were tiny particles that traveled through the vacuum of space, or space was not a vacuum but contained some mysterious substance that carried these forces. Thus began the wave-particle debate. A substance requires a wave; a vacuum requires particles.

In the beginning, the particle view was largely accepted, but by the nineteenth century the wave view predominated, and along with it the concept of the ether was born. The ether (an idea first suggested by Aristotle) was thought to be a motionless fluid that permeates space and carries electromagnetic waves. This conception requires that light be a transverse wave, which means that the distortion of the ether is at right angles to the motion of the wave. Since the rate at which a transverse wave moves depends on the force that restores the distorted region, and since light travels at an extremely rapid velocity, the force holding the ether in place would have to be enormous. In fact, it was calculated to be more rigid than steel. On the other hand, the ether had to be a medium that did not interfere with the motion of heavenly bodies in any way that could be observed. So the ether was somehow required to be stronger than steel yet be a rarified gas. One can see why visualizing it caused so many problems.

Considering the ether as the material through which the force of gravitation operated also had difficulties. Light can be reflected and refracted,[4] but gravity waves cannot. If the ether carried both, why the difference? (The problem of two separate ethers was not addressed, but just in case a second ether might be proposed for gravity, it became fashionable to speak of the "luminiferous" ether as the one carrying light.) In spite of these problems, classical physicists accepted the concept of the ether, even though it violated common sense. But, as we shall see, common sense is often contradicted by physical theories.

Attempts by nineteenth-century scientists to detect the ether were based on the assumption that the air surrounding the earth is analogous to the ether that fills the universe. That is, both are material substances made of particles. If the ether does not move as the earth orbits around the sun, it was reasoned, then the earth must be moving relative to the ether, just as a runner moves relative to the air. If the air is motionless, the runner will feel a wind in her face; in an analogous manner, the movement of the earth should create an "ether wind." With the earth moving around the sun at about thirty kilometers per second, and adding to that the earth's rotation and perhaps the motion of the sun, quite an ether wind should blow through the physicist's measuring equipment. On the other hand, since the ether offers no impediment to moving bodies, no ether wind should be detectable.

If, however, the ether is assumed to carry the waves of light, another experimental approach is possible — detecting the ether by

---

[4]Refraction is the change of direction a wave undergoes when it travels from one transparent medium to another. Today we know that *in principle* gravity waves could be reflected and refracted, though it takes enormous curvatures to get such an effect.

measuring the velocity of light in opposite directions. This idea led to the famous experiment conducted by Albert Michelson and Edward Morley.

The Michelson-Morley experiment and the Lorentz-FitzGerald contraction

Imagine a measuring instrument sitting stationary on the earth with the earth moving around the sun at thirty kilometers per second. If a beam of light comes directly towards the instrument and if both the beam of light and the instrument are moving in the same direction relative to the ether, the velocity should measure less than if the light comes from the opposite direction. In 1881, Michelson and Morley developed equipment that could detect this small difference in velocity.[5] But their results were negative: they found no variation in the velocity of light. The two experimenters made thousands of observations, always with the same outcome. By 1887 they were forced to conclude that there is no ether wind. The same experiment has been repeated many times since, and the Michelson-Morley conclusion has been consistently confirmed. The implication is that there is no way of determining absolute motion through space — or, put another way, that the velocity of light is constant and independent of the observer. To nineteenth-century physicists, this seemed preposterous.

Among suggestions made to save the ether concept, the best was offered by George FitzGerald in 1893. FitzGerald's theory proposed that all matter contracts in the direction of motion because of the pressure exerted by the ether wind. The degree of contraction

[5]The Michelson-Morley experiment splits a single beam of light so that one part goes off at right angles to the other. The two parts are then reflected by mirrors equidistant from the split and returned to the same point. If the first beam is going with the ether wind, the second beam is going crosswind. If the source of light is at rest with respect to the ether, then the two light beams should return to the same point at the same time. However, if the light source is moving with respect to the ether, then an interference pattern will be noticed when the two beams come together.

would increase with velocity since the pressure of the ether wind would increase. Such contraction would explain the negative results encountered by Michelson and Morley. FitzGerald came to this conclusion by assuming that electromagnetic forces hold matter together. This led him to surmise that a material body would contract in the direction of motion if the velocity of the body approached that of light. But this contraction would be unobservable since the measuring devices, also subject to the earth's motion, would also contract. In this way, FitzGerald was able to explain why light has a constant velocity regardless of the motion of the observer or the source. The theoretical distortion of the measuring instruments — small at everyday velocities but increasing rapidly at high velocities — was the first indication that the velocity of light might have deep significance.[6]

In FitzGerald's scheme, as an object approaches the velocity of light, its length approaches zero. If the velocity of light is exceeded, the length becomes an imaginary number, a mathematical entity with no physical meaning. (We will deal with imaginary numbers later.) Hendrik Lorentz extended FitzGerald's explanation by discovering that when a particle carries a charge, not only does the length contract but the mass increases with motion. While this effect is negligible at ordinary velocities, it shows up dramatically when the velocity approaches that of light. At the speed of light, the mass becomes infinite. Again, the velocity of light becomes an immovable barrier. While the Lorentz-FitzGerald contraction, as it came to be known, offered an ingenious explanation for the results of the Michelson-Morley experiments, one fact remained abundantly clear, at least to Einstein: it was impossible to measure absolute motion.

---

[6]The existence of the ether wind would make one of the two beams described in footnote 5 take a longer time to return. But the FitzGerald contraction allows the delayed beam to complete its journey in the same time as it would if there were no ether wind. The two effects cancel beautifully.

This point deserves some elaboration. If one imagines that the earth is at rest, the stars seem to be moving around the earth. But one could describe the earth as turning with the stars standing still. Originally, scientists assumed that the earth is at rest since we cannot perceive its motion, but as knowledge of astronomy developed, it became clear that the earth is turning and not the stars. Likewise, the sun, around which the earth revolves, turned out to be a moving star. So, by the middle of the nineteenth century, it was assumed that no material object could be at absolute rest. This information cast a shadow on the validity of Newton's laws of motion, based as they were on notions of straight lines and absolute rest. That is, if two observers were moving relative to each other, whose perception would be correct? Postulation of an all-pervading ether resolved this problem by allowing absolute motion to be established. As the Michelson-Morley experiments showed, however, absolute motion could not be measured because of the Lorentz-FitzGerald contraction. It is important to note that nineteenth-century physicists did not say that there is no absolute motion, only that it cannot be measured. To sum up, by 1900, space was envisioned as filled with an ether that was not "real" like other substances in the universe since it could not be detected by any means then available.

Matter and fields

While space was something of a puzzle to the classical physicist, there were few doubts about matter. Perceptible to anyone endowed with normal senses, matter is the constituent substance of everything from people to the earth to clouds and stars. Matter is measured in terms of mass, of which there are two types. The inertial mass of a body is its tendency to resist a change in motion, and the gravitational mass is its attraction to other matter. The idea that mass is conserved[7] regardless of the changes going on in a system was accepted until Einstein published his relativity theory.

---

[7]A conservation law states that some property of a system remains constant through change.

While the world is composed of matter, it also contains fields. These fields are of two types, electromagnetic and gravitational. Objects with an electromagnetic charge appear to attract each other (or repel each other in the case of the same charges) by forces that move across space. The force of gravity is similar in that it acts at a distance. So, there is a new kind of geometric unity that Einstein sought, in that all force fields are curvatures modifying some kind of space. Gravity modifies space-time. The electromagnetic field modifies a new kind of space beyond, but still attached to, space-time.

It was James Clerk Maxwell who placed the phenomenon of the electromagnetic field on a firm mathematical basis. By the mid-1800s, the relationship between electricity and magnetism was well established. (The motor and generator were already in use.) In 1864, Maxwell developed four equations which described the integration of the magnetic and electric fields. The results were that a changing electric field induced a changing magnetic field, which returned the favor by inducing a changing electric field again. This procedure produces a field that radiates outward in all directions, and this radiation is wavelike. The electric and magnetic components are at right angles to each other and to the direction in which the wave is propagated. In effect, Maxwell showed that the electromagnetic field is a separate entity in that it can sustain itself without an electric charge or current. Further, if a test charge is placed in its path, the charge recoils because the wave imparts some of its momentum to it. So, the electromagnetic wave is energetic in that it carries momentum and can exert a force on matter. When Maxwell calculated the velocity of these waves, it turned out to be dependent on the electric and magnetic forces between charges and between magnetic poles. In a vacuum, the velocity was equivalent to the velocity of light. From this Maxwell deduced that light must be electromagnetic radiation.

(Maxwell, being a product of his times, attributed the oscillations of this field to distortions in the ether.) Maxwell's equations also predicted electromagnetic radiation at frequencies other than those for visible light, frequencies which were discovered later by such physicists as Heinrich Hertz and Wilhelm Röntgen.

As Einstein pointed out, the classical physicists made extraordinary attempts to save the mechanistic point of view. Various artificial substances and concepts like electric and magnetic fluids and light corpuscles were introduced to accommodate the anomalies and difficulties. But the ether never satisfactorily explained optical phenomena. Construing the ether as material consolidated many of the difficulties faced in classical physics, but in the long run it was impossible to paper over the immense problems created by this view. Thus most physicists came to recognize that the mechanistic standpoint does not suffice to explain all the events of nature.

As we end our historical excursion through classical physics, we are left with two concepts to describe reality: matter and fields. It seems natural to assume that matter and fields are aspects of the same thing. Indeed, this assumption has been convincingly demonstrated in the past twenty years in what may be called the standard model of elementary particles.

To summarize the classical view:

- Newton's three laws of motion provided the foundation for a mechanistic and deterministic view of the universe.
- The first law defines the principle of inertia and notes that a mass will continue being at rest or in uniform motion if it is not acted upon by an external force, suggesting that matter is inert, having no inherent power of action.
- Since the three laws were not sufficient to cover such phenomena as falling apples, Newton added the law of gravitational force.
- Since material bodies seem to interact without engaging in

physical contact, the problem of action at a distance arose. This interaction had to explain action across a vacuum, leading to the possibility that the vacuum is not empty but contains a non-material substance called the ether. Action at a distance could be explained either by waves in the ether or by tiny invisible particles which carry the action. Thus began the wave-particle dichotomy.

- ☐ The Michelson-Morley experiment led to the conclusion that absolute motion cannot be established experimentally.
- ☐ Maxwell established that electromagnetic energy could be transmitted from one point to another by a mysterious entity called fields; thus, even in the classical view, the world consisted of something beyond mere matter in motion.

2 The contemporary view

One remarkable thing about physics in the twentieth century is that two major scientific revolutions took place almost simultaneously. They are represented, of course, by the theory of relativity and the quantum theory. Another remarkable thing is that the conceptual and mathematical implications of these theories cannot be reconciled into a single theory because some of their basic axioms are contradictory. For example, relativity requires continuous space-time, strict determinism, and locality, while quantum theory involves discontinuity, indeterminism, and nonlocality. (Locality is the principle that defines the causal relationship between events — that is, in a "local" reality, information cannot travel faster than light — whereas nonlocality refers to instantaneous action at a distance.) Some aspects of relativity and quantum theory have been reconciled, but largely the quest for theoretical unity has been elusive.

If relativity and quantum theories clash in their basic assumptions and if each provides a partial explanation of our experience, it is logical to assume that both theories are portions of a broader view that should explain more phenomena than the two theories do separately. Also, it is interesting to note that while both theories seem to have basic assumptions that are in direct opposition, they have one common characteristic: both point to the concept of wholeness, though in different ways. Here are the comments of two physicists regarding relativity theory, quantum theory, and wholeness. First, Mendel Sachs in reference to relativity:

> . . . the theory of relativity, when expressed in its full form . . . [implies] that the world is continuously one — a closed system of inseparable components. (Sachs, p. xi)

Second, David Bohm in reference to quantum theory:

> One is led to a new notion of unbroken wholeness which denies the classical analyzability of the world into separately and independently existing parts . . . . The inseparable quantum interconnectedness of the whole universe is the fundamental reality. (Herbert, *Quantum Reality*, p. 18)

Wholeness, as we shall see, is essential to any theory that can embrace both relativity and quantum theories.

Relativity theory

The special theory of relativity

The special theory of relativity[1] is based on two assumptions. First, while all motion can be seen as relative to some object, no object can be considered at absolute rest. The Lorentz and FitzGerald discovery that the motion of an object cannot be measured to equal or exceed the velocity of light led to Einstein's second assumption: that the velocity of light in a vacuum will always be the same regardless of the motion of the source or the observer. That is, if you attempted to catch a pulse of light via a rocket ship, you would never gain on the pulse no matter how fast your rocket. The significance of this velocity-of-light barrier is profound, for it requires us to give up the simple additions of velocities, which is almost like saying three plus three does not equal six. Nevertheless, the constancy of the velocity of light regardless of the motion of the observer or the source is a fundamental law of nature.

The fallacy in the simple addition of velocities, as we shall

[1]Though relativity theory presented radical modifications to classical physics, it did not question the concept of causality as quantum theory did. Since according to relativity theory systems evolve deterministically, its formulation is frequently considered to be an extension of classical theory. The radical departure of Einstein's theory from Newtonian mechanics is in asserting fundamental principles regarding energy and matter, space and time, and the nature of light that are utterly divergent from our normal experience.

see, is that distance and time are relative concepts dependent on the motion of the observer. Space is not constant: if you travel fast, distance seems to shrink. It is important to remember that Einstein did not say a body could not exceed the velocity of light but that a body could not be *accelerated* to the velocity of light without the application of an infinite force. Therefore, if a body were created with a velocity greater than light, this would not be a violation of the theory. Furthermore, if a body were traveling at a superluminal (faster than light) velocity, we could not measure its velocity nor would we be aware of its existence.[2] To Einstein this was tantamount to saying that no such body exists (at least as far as the physicist is concerned).

Returning to Einstein's first assumption regarding motion as relative, if a particle moves uniformly through space,[3] its velocity cannot be ascertained by examining effects within the particle, only by considering its surroundings. To Einstein, that meant it was impossible to measure absolute motion. This concept rejected Newton's notion of an absolute space, at rest and immovable, relative to which all objects in the universe are moving. Einstein suggested that this description of space be discarded because any observer, in whatever frame of reference, could say that he or she was at rest and all else was moving relative to them. This in effect removed any system from being a special frame of reference.

Einstein's second assumption was that the velocity of light in a vacuum is independent of the motion of the observer or the source. In our ordinary world (at velocities well below that of light) this notion seems counterintuitive and also violates Newton's laws. That

[2]It is true that if the tachyon (discussed in Chapter 11) is charged, one could see Cerenkov radiation, which is the excess energy that results from the difference in velocity between the particle and its associated electric and magnetic fields. These fields cannot exceed the velocity of light in the medium. An analogous concept is the so-called sonic boom.

[3]Einstein's special theory applies only to inertial frames of reference moving uniformly with respect to each other.

is, Newton's laws predict that light travels at different velocities for observers moving at different speeds. For example, suppose two people are throwing a ball back and forth on the deck of a ship. Let us assume that the ball is being thrown at six miles per hour and the ship is moving at twenty miles per hour. If the ball is thrown in the direction the ship is moving, its velocity relative to the ocean is twenty-six miles per hour. On the return throw, the velocity of the ball relative to the ocean is fourteen miles per hour. This calculation (resulting from Galilean relativity) amounts to a simple addition and subtraction of velocities, all of which makes intuitive sense. But suppose we are throwing a beam of light rather than a ball, and suppose the ship's velocity is approaching that of light. An observer standing out on the ocean would find the velocity of the thrown light to always be the same, and independent of the velocity of the ship. That violates Newton's laws but is correct according to all experiments ever performed to check this phenomenon. A corollary of this assumption is that any material body will always be measured at a velocity less than the velocity of light.

Even though Einstein's name is synonymous with relativity, the concept of relative motion was not new with his theory. His great contribution was in recognizing that the velocity of light is *not* relative — it is absolute and it is the only absolute motion in our universe (it is the frames of reference that are relative). In fact, Einstein preferred the name "theory of invariants" to "theory of relativity" for his discovery.

Perhaps an analogy will help explain how light, as an electromagnetic wave, is invariant and is different from a mechanical wave such as water. Picture yourself in a boat with a wave moving across the surface of the water. If your boat travels in the same direction as the wave and at the same speed, the wave will seem stationary relative to the boat. If you increase the speed of the boat, you can outdistance the wave. This is not true with light waves. Whereas the speed of mechanical waves is determined by the characteristics of the medium (in this case, water), light waves are electromagnetic and can exist in empty space (without a medium).

A constant velocity for light regardless of the motion of the observer or the source does not violate Maxwell's laws, but does, as we said, contradict Newton's. As pointed out earlier, according to Maxwell, the speed of light is constant and is determined by the ratio of electric and magnetic forces. But Newton said the velocity of light would be dependent on the direction of the earth's motion through the stationary ether-sea. In his development of relativistic mechanics, Einstein accepted Maxwell's idea and rejected Newton's. The resulting formulation changed physicists' understanding of space and time in dramatic ways.

The velocity of light, which results from a varying electric field generating a magnetic field and a magnetic field generating an electric field, is an intrinsic and essential component of the universe, and is one of the fundamental constants of nature which never varies. It is not like the velocity of a pitched ball which depends on the direction and force applied by the thrower; it is always and invariably the same. The velocity of light and a number of other constants (such as the gravitational constant and Planck's constant, which will be discussed later), the electric charge of the electron and proton, and the masses of some subatomic particles, all define our universe in very specific and precise ways. Even small changes in these constants would cause drastic changes in the structure of our world.

Einstein's mathematical formulations describe how space, time, and mass change in a moving frame of reference. That is, if the velocity of light is constant, each observer's perceptions of space and time must differ depending on their state of motion. As we shall see in Chapter 4, what we call space and time are really projections from a space-time world, and they vary according to the velocity of the reference frame of the observer, so that the distance between two points is different for different observers. This astonishing conclusion means that space is no longer considered a rigid volume in which objects can be located, but has an elastic quality.

Each observer can use Einstein's equations to calculate exactly what a second observer will measure. Equally amazing, the equations can be reduced to the Lorentz-FitzGerald contraction and Lorentz's gain of mass with motion. Lorentz's conclusions applied only to charged particles, but Einstein showed that the phenomenon could be applied to all objects. In discarding an absolute space and a uniform flow of time for all observers, Einstein was required to introduce a new absolute — the velocity for passing information in our three-dimensional world. That is, all observations have a built-in delay, which results in the Lorentz-FitzGerald anomalies.[4]

The finite and maximum velocity of light leads directly to Einstein's famous formula, $E=mc^2$, which states that energy and mass are equivalent. The conversion factor is the square of the velocity of light. In classical physics, mass and energy are governed by separate conservation laws;[5] relativity theory removed the distinction between the two. In the classical view, a body at rest has mass but not kinetic energy. If it is moving, it has both mass and kinetic energy.[6]

---

[4]One could ask if the length of a yardstick traveling at near the speed of light would contract for a stationary observer, since for an observer traveling alongside the yardstick, no contraction takes place. Whose observation can be considered real since an absolute space has been discarded? As we have seen, the Lorentz-FitzGerald contraction is only an apparent effect of motion of an object relative to an observer. This anomaly suggests that spatial and temporal concepts are modes by which we think, not concepts independent of the observer.

[5]Since the classical physicist viewed mass and energy as separate properties, physics before relativity had two conservation laws rather than one.

[6]The kinetic energy of an object is defined as work it can do because of the motion of the object. For example, the work a medicine ball can do in knocking down a building is its kinetic energy.

Suppose a body moving along a straight line is acted upon by an external force that increases its velocity from 100 miles per second to 120 miles per second. From the classical point of view, that force always produces the same change in velocity in the same time regardless of the initial velocity. But we know from the Lorentz-FitzGerald contraction that if the initial velocity is close to the speed of light, the same force will not produce the same change in velocity. A much larger force is required. Since mass is defined by its resistance to a change of velocity, we may ask, what mass is resisting the force at the higher initial velocity that is not there at the lower velocity? It turns out that the additional mass *is* the kinetic energy, which increases with velocity. This equivalence of mass and energy, however, was not known before Einstein because a given amount of energy represents such a proportionately small mass.

Einstein's mass-energy equation certainly put a strain on the classical physicist's view of matter as a substance. Instead of being quantified by its mass, matter now must be quantified by both its mass and energy. But energy in turn is dependent on the speed of matter, which is dependent on the motion of an observer. For example, if you are in a car going sixty miles per hour, then to you the energy of motion of the car is zero. But to someone sitting in the middle of the highway, the energy of the car is quite different. So the measure of matter (mass and energy) is observer-dependent, and the whole notion of matter becomes somewhat hazy.

Since Einstein assumed that it was not possible to measure absolute motion, it is commonly believed that he discarded the notion of an ether. What he actually said was that the ether could not be constituted of moving particles. Put another way, he redefined the ether as space-time. With water waves, particles of water change their position with time as the wave passes through, and this motion can be tracked by using floats (such as corks). If the motion of water particles could not be tracked, then the assumption that the water is composed of particles would be unfounded, but we could still view the water as a medium. Similarly, Einstein did not see space-time as pure nothingness nor as a lattice of ether atoms, but as a continuous, geometric jelly that could be stretched and distorted; it could have bumps, which could not move in an absolute sense but could move relative to each other.

The general theory of relativity

This idea brings us to Einstein's general theory of relativity. In developing this theory, Einstein used a metaphor suggested by Hermann Minkowski, one of his former teachers. In 1908, Minkowski introduced the idea of space-time,[7] in which space and time were no longer seen as independent. Minkowski saw Einstein's new laws (in special relativity) as describing objects in four-dimensional space-time, with three dimensions of space and one of time. This is not a Euclidean space for two reasons. First, for time to be expressed as space, time must be multiplied by the velocity of light. Second, the time dimension is imaginary, which in special relativity creates peculiarities that will be described later in some detail (pp. 67–68).

To visualize this four-dimensional space-time, imagine a plane with a horizontal coordinate labeled "space" and a vertical coordinate labeled "time." Suppose you are at rest at point zero, which is where the space and time axes cross. If you do not move, you have no component along the space axis. But along the time axis we

---

[7]Minkowski unified space and time, but in such a way that timelike and spacelike directions can be distinguished. To Minkowski, the Lorentz transformations, which defined shifts in perspective between two coordinate systems traveling at different velocities, turned out to be a simple rotation of the space-time axis. A metaphor often used is that these rotations and the resulting contractions of rods and dilations of clocks are different slices of a sausage. The sausage is always the same and objective, but by altering the direction of the slice, one obtains different shapes. Some physicists see the Minkowski universe as a static one since the sausage remains the same.

have a different story: time does not rest but continually moves on. If the time axis is spatialized, its distance measurement is the velocity of light multiplied by time. Since time is moving, at rest you are moving down the time axis at the speed of light (300,000 km/sec[8]). This is possible because the speed of light is constant for all observers. The reason you do not perceive this movement is that the entire universe is engaged in this process, and the time axis does not allow free movement as does the space axis. Another way of stating it is that 300,000 km/sec is a manifestation of one second of time — and one second is defined as a distance of 300,000 km. This way of looking at it allows us to see time as a fourth dimension.[9]

As we pointed out, for every second on the time axis, we have "traveled" 300,000 km. In Minkowski space-time, things are replaced by events and events become points. So while measurements of length, clock rates, and mass seem to vary when an observer changes his or her state of motion, Einstein's four-dimensional objects do not change their form when the observer moves around. That is, when space contracts, the time changes in just the right amount so the total effect on all four dimensions remains unchanged. Thus the laws of physics become invariant for all observers.

The peculiar four-dimensional world that Minkowski describes is not at all the world of our experience! Space-time is a changeless state in which past, present, and future events exist

---

[8]We are using motion here in a metaphoric sense. For example, in his excellent book *Parallel Universes*, Fred Alan Wolf says,

> As we sit, we say time marches on. Actually time doesn't do that. We march on in time. In fact, from the point of view we are taking here, we march on through an imaginary dimension of space called time, moving at the speed of light. (Wolf, p. 126)

[9]As we imagine our universe moving down a fourth dimension we call time, the velocity of light is in some sense equivalent to the velocity of time. We shall see later that our velocity of light defines our time, but that other universes with other velocities of light might experience time differently.

simultaneously. In the same way that space is just there, so is time. Space-time does not endure or continue in time, just as it is not spread in space. Time exists only because as conscious beings we seem to require it. Thus, consciousness seems to enter Minkowski's world and is related to the creation of time.

To deal with this four-dimensional space-time, Einstein used a new kind of geometry developed by mathematician Georg Riemann, which allows an intrinsic curvature that can be completely measured by observers confined to four-dimensional space-time. (Technically, curved space means that Euclidean geometry does not hold.) From this concept Einstein developed his general theory of relativity. The special theory of relativity was limited to the observations made from frames of reference moving relative to each other with a constant velocity. In the general theory, Einstein investigated the effect of relative accelerations in contrast to relative uniform velocities and found a relationship between gravitational effects and curved space-time. More specifically, he interpreted the gravitational force as the bending of space-time.

According to Newton's law, gravity affects all matter in an equal manner. That is, the force of gravity does not care what the internal structure or the mass is for a given body. Since all bodies accelerate equally under gravity, one could assume that gravity is a function of the surrounding space, rather than a function of the make-up of the bodies in motion through space. From this notion Einstein developed the idea of gravity as a distortion in the geometry of space-time. He described this in detail, with the help of Riemann's mathematics, in a set of ten equations called field equations, which make up the general theory of relativity. These field equations indicate the connection between matter and space-time geometry. However, where the curvature is not strong, the equations are the same as the classical Newtonian equations.

If you ignore air resistance, two objects of different mass dropped together from a great height will travel downward at the same acceleration. If you happened to be sitting on one of those objects and looking at the other one, it would appear to you as if the second object were suspended in space in close proximity to you — as if the law of gravity were suspended. This occurs when astronauts take a space walk: in free-fall along with the spaceship around the earth, they appear to defy gravity. That is similar to the moon "falling" around the earth: the moon does not fall into the earth because it is falling away from a straight line, which is the path it would take if no force were present. Physicists conclude that the effects of a gravitational field are similar to those of acceleration. Metaphorically, this concept is often illustrated by an elevator. If the elevator is accelerated upward, passengers inside feel pulled to the floor. But from their vantage point, they cannot determine whether the elevator is accelerating upward or they are being held on the floor by gravity.[10]

In his general theory, Einstein discarded the notion of gravitational force just as he did the notion of a material ether. Whereas Newton gave rules on how bodies react to forces between them in a standard Euclidean space, Einstein said that bodies at rest or in motion change the local topography of space-time. Each body follows the path of least resistance through all the distortions of space-time created by all the bodies in the universe, so a force between bodies is not necessary. What is necessary is the non-Euclidean geometry of Einstein and Riemann.

The best way to understand the similarity between geometric curvature and gravity is to visualize a large rubber sheet stretched flat with a small ball moving along the sheet at a constant velocity.

---

[10]This concept is known as the equivalence principle of relativity theory. That is, in an experiment, a physicist cannot determine whether one is encountering a steady acceleration or gravitational attraction.

The ball will obey Newton's first law of motion and follow a straight line at a constant speed. Now imagine a heavy object in the center of the sheet, with sufficient mass to depress the rubber sheet around it, but leaving the sheet flat some distance from the depression. As the ball approaches the depression, its initial speed changes and it can be made to follow an orbit around the depression, which mimics the properties of gravity described by Newton. On a flat rubber sheet, the ball seems to move inertially, and if the sheet is curved, the ball moves as if subjected to a gravitational force.

When Einstein discarded the Euclidean geometry of the Newtonian universe, he replaced gravitational forces with a space-time that had a varying curvature. A curved orbit in Newton's world would be a straight orbit in the curved space-time of Einstein's world. This straight line in a curved space is called a geodesic. To an observer from a different frame of reference, the geodesic may be curved; for example, the geodesic on the surface of a sphere is a great circle. In general relativity all inertial or free-fall motion moves along a geodesic. Since we are dealing with space-time as a single entity, gravity also affects time.[11] An observer distant from the vicinity of a large mass will notice a slowing of time in the area of the intense gravitational field created there. In other words, gravity slows time for an outside observer.

From all we have said, it can be seen that matter — the ball on the rubber sheet — causes strains in space-time and thus creates curvatures. The degree of curvature is proportional to the mass. Referring back to the special theory of relativity, we know that mass is equivalent to energy, so light also is deflected in a gravitational field. In fact, Arthur Eddington verified Einstein's predictions by measuring the deflection of light by the gravitational field of the sun.

---

[11]Astronomers have observed that the spectral lines (created when atoms make definite transitions between energy levels) of certain celestial bodies seem to have

He did this in 1919, when he led a scientific expedition to Principe Island off the coast of Spanish Guinea. There he was successful in photographing the deflection of light around the huge mass of the sun. To accomplish this feat, he had to observe the sun during a total eclipse. The stars that were at a tiny angle to the sun were seen to be out of position by a very small amount. While the deflection was small, it matched Einstein's prediction that light does not travel through space in straight lines but follows the curves of space-time.

Einstein's field equations have very few exact solutions. In Newtonian physics, when a body creates a gravitational field, it does so everywhere instantaneously. Not so in Einstein's physics: the space-time curves we have been discussing travel at the speed of light. Space-time distortions contain energy, and when these distortions move, waves travel out at the speed of light. But these "gravity waves" are very weak and difficult to detect.[12]

Einstein felt that the gravitational field must be only one aspect of some overall field. If this were true, geometry could somehow describe the properties of matter. Toward this goal, Einstein attempted to unify electromagnetism and gravity, but without success. A unification of space-time and matter could be attempted by materializing space-time or geometrizing matter. Einstein chose the second option. If that turns out to be correct, then physics at its most basic is essentially the study of the geometry of space-time, which would imply that space-time and matter are a continuum that cannot be represented accurately by separable parts. Classically,

slightly larger wavelengths than normal. That is, the spectral lines seem to shift toward the red end of the visible spectrum. Sometimes this red shift indicates the object is moving away from the observer, in which case it is called the Doppler effect. However, the red shift can also be caused by a high gravitational field which produces a time distortion.

[12]Recent studies indicate that if two dense stars orbit about each other in a system known as a binary pulsar, the movement creates gravitational energy in the form of waves, which alters the orbital period. The detection of these gravitational waves confirms the predictions of general relativity.

elements of matter are fundamental units whether they be atoms, elementary particles, or whatever. On the other hand, relativity suggests that matter is a manifestation of space-time — somewhat like ripples on the surface of a pond — and the two basically are not separable.[13]

In discussing concepts of relativity that seem counterintuitive, it is useful to employ metaphors and analogies. To help us picture the unpicturable, let us turn to the comments of William Clifford and John Wheeler. As a young man, Clifford translated the work of Riemann, whose theories Einstein used in relation to the curvature of space-time. Clifford, who also was impressed with Riemann's work, noted that the curvature of space might be nearly uniform but with variations from point to point. Further, these local curvatures could vary with time.

Some physicists understood these curvatures to be caused by unseen forces and to be independent of the geometry of the space, but Clifford and Einstein saw the geometrical structure as their source and hypothesized an additional dimension for the geometry to curve into. The curvature is like the indentation in the rubber sheet in the example described above — but without the ball. Rather, the curvature itself, that is, the distortion of space-time, is what gives rise to matter. In Clifford's words, matter is bumps in space that vary in position with time. In the late 1870s, he published the following assertions:

---

[13]Here is the way the physicist Jack Sarfatti states it:

> In fact, this vision of Einstein's has been fulfilled in recent years in the standard model of elementary particles and fundamental forces. Einstein's original four-dimensional space-time is now understood to be a projection, or a shadow, of a larger higher-dimensional space called the "fiber bundle." The elementary particles and nongravitational forces are then interpreted as geometric structures in the higher dimensions beyond Einstein's space-time. (Sarfatti, private communication)

> (i) That small portions of space *are* in fact of a nature analogous to little hills on a surface which is on the average flat; namely, that the ordinary laws of geometry are not valid in them.
> (ii) That this property of being curved or distorted is continually being passed on from one portion of space to another after the manner of a wave.
> (iii) That this variation of the curvature of space is what really happens in that phenomenon which we call the *motion of matter*. . . .
> (iv) That in the physical world nothing else takes place but this variation, subject (possibly) to the law of continuity. (Weaver, Volume III, pp. 680-681)

Clifford died the year Einstein was born, having stated the essence of relativity theory forty years before publication of the special theory.

Building on the work of Riemann, Clifford, and Einstein, Wheeler developed the concept of geometrodynamics. We shall investigate his ideas more fully in connection with quantum theory, but for now, suffice it to say that to Wheeler, the four-dimensional geometry we have been discussing is all there is "out there." He speculates that a slow curvature might be a gravitational field, a rippled geometry might be an electromagnetic field, and knotted moving regions of space might be particles. With this notion, matter is not a separate, independent entity but is an aspect of the geometry of space-time. As Wheeler describes it,

> There is nothing in the world except empty curved space. Matter, charge, electromagnetism, and other fields are only manifestations of the bending of space. Physics is geometry. (Weaver, Volume II, p. 263)

Now let us develop a metaphor for describing physics as geometry. Again imagine a rubber sheet, but now assume that it forms a sphere. On the surface of the sphere is a population of two-dimensional beings, the counterparts to our three-dimensional selves. These flat beings cannot imagine anything beyond the surface of their sphere (just as we live on a sphere but in our everyday activities consider the earth a flat plane). All of the universe exists there on the surface; all travel from one place to another takes place on the surface. A straight line is a great circle. But let us assume that the science of these beings is advanced (they have their own Einstein), and, while they cannot picture what it would be like to be above the sphere or in its interior, they are aware that their space seems to be curved. So it is with us. Although it is difficult to imagine time as the fourth dimension, nevertheless, our science tells us our three-dimensional space is curved into that fourth dimension.

According to Newton's law, gravity obeys the principle of superposition, meaning that the gravitational force on a given body can be the sum of the forces of several different bodies existing at different parts of space at the same time. In Einstein's theory this does not hold exactly in that two different patterns of distortion in space cannot be added to arrive at the total distortion. In some complicated way, the distortions that the patterns produce also act on each other. This is called self-interaction, and it indicates that curvatures of space-time are also a form of energy, which produces its own gravitational field. To be noticed easily, this effect requires an enormous gravitational field, but the point is that all curvatures, large or small, are forms of energy, so not only matter but electromagnetic and gravitational fields are energetic. That is, on the sphere made of the smooth rubber sheet, all distortions of the surface contain energy. This information will be important when we examine matter fields in quantum theory.

Summary

To sum up this discussion of relativity theory, let us reiterate that it requires a continuous space-time, strict determinism, and locality. Important points to remember are:

- According to Einstein's theory, space-time does not have separate, discrete parts that can be tracked through time but is a continuous, nonmaterial jelly that can be stretched and distorted, with bumps which move about relative to each other.
- In Minkowski's four-dimensional space-time, past, present, and future are equally real, implying a strict determinism.
- Finally, until very recently, it was generally believed that all the curved deformations of space-time are not instantaneous but travel at the velocity of light, and therefore locality holds. (As you recall, locality means no information or influence can travel faster than light.)[14]

Though it maintains determinism and locality, relativity does imply wholeness: space-time is one complete continuum.

Quantum theory

As we have seen, relativity theory is concerned with the structure of space-time on the large scale and with the way matter affects the geometry of that structure. Quantum theory deals largely with matter on the small scale of elementary particles and atoms, but it is a consistent theory which applies also on the macro level, though its effects there are small. For example, such widely disparate issues as the solidity of bodies and the underpinnings of chemistry are all explained by quantum theory.

The first way in which the basic assumptions of quantum theory differ from those of relativity theory is that relativity requires continuity while quantum theory is grounded in discreteness. Discreteness means that a system changing from one state to another does not necessarily go through a series of intermediate states which are similar to the initial and final states; that is, an electron can go from state A to state B without traversing the space in between. This

---

[14]Recently, some physicists have been questioning this statement, feeling that it is not supported mathematically.

motion, known as quantum jumping, was the cause of many a sleepless night for physicists in the 1920s.

The concept of discreteness was introduced in 1900 by Max Planck in an attempt to solve a problem related to black-body radiation. A black body is defined as an object that does not reflect light of any frequency. Since it absorbs all frequencies, when the body is brought to incandescence, it should emit all frequencies. Planck found that the theory in vogue at the time was unable to explain the relationship between the temperature of the black body and the amount of electromagnetic radiation emitted at various frequencies. Though this theory predicted fairly accurately the amount of energy emitted at low frequencies, at high frequencies it predicted the emission of an infinite amount of radiation. This phenomenon was called the ultraviolet catastrophe and certainly did not agree with experimental results, which showed that radiation was low at both the low and high frequencies and peaked in the middle. To fit the experimental curve, Planck proposed that the probability of radiation decreased as the frequency increased. This required accepting the fact that energy does not flow continuously but is radiated off in discrete quantities. These quantities he called "quanta" and thus quantum theory was born. Not only did energy become "atomized," but the energy of each particle increased with frequency.

The proportionality constant required to relate energy and frequency is called Planck's constant, designated by the letter *h*, and is equal to 6.6×10^{-34} joule seconds. (The joule has replaced the calorie as a unit of heat energy; one calorie is equal to 4.19 joules.) This is so small that the quantum of energy was never observed by physicists. While Planck's discovery was little noted at the time, in hindsight we can see it as the demarcation point between classical physics and contemporary physics.

One person who did take notice of Planck's discovery was Einstein. In 1905, Einstein suggested a solution to the problem of the photoelectric effect, namely, that electrons are emitted from a metal only after a certain threshold frequency is reached. Further, an increase in *energy* of the light impinging on the surface of the metal does not increase the kinetic energy of the emitted electron, but if the *frequency* of the light is increased, the kinetic energy goes up. Classical physics could offer no explanation for this, but according to the formulas developed by Planck, it can be explained if light is seen as a stream of light quanta rather than waves. Einstein started with Planck's discovery, then assumed that light was absorbed in quanta as well as radiated, and went on to perfectly explain the photoelectric effect. Now the physics community took notice of the quantum idea.

Even though Einstein utilized the idea of light quanta, he and other physicists were still reluctant to face the fact that light could be both a wave *and* a particle. Most of them viewed the light quantum as necessary mathematically but without physical significance. By the 1920s, the concept was here to stay, and in 1926 Gilbert Lewis named the light quantum the photon. Einstein never came to terms with this. In 1951 he remarked:

> All these fifty years of pondering have not brought me any closer to answering the question, What are light quanta? Nowadays every Tom, Dick and Harry thinks he knows it, but he is mistaken. (Coveney and Highfield, p. 114)

Planck and Einstein were due a lot of credit for solving the problems of black-body radiation and the photoelectric effect, but it was something else that made the quantum idea revolutionary. For about one hundred years, light had been considered a wave, a concept that brilliantly explained the phenomenon of interference, in which waves that are out of phase create a characteristic pattern. If light is seen as particles, interference cannot be explained. The only conceivable answer to this problem shocked and baffled physicists of the time: light would have to be both a particle and a wave. In this way, physicists were led from the notion of discreteness to acceptance of the duality of energy. That is, light can behave like a wave or a particle depending on the experimental arrangement — one of many strange features of quantum theory.

In 1913 Niels Bohr suggested a new model for the atom that incorporated the budding quantum theory. One previous model, proposed in 1898 by William Thomson, was a sphere with electrons embedded within it. The sphere was positively charged with just enough electrons (negatively charged) to make the atom neutral. A major drawback to this concept was that by the early years of the twentieth century, physicists had concluded that the atom is not a solid mass but mostly empty space. In 1911 Ernest Rutherford proposed the now-famous planetary model consisting of a nucleus with a positive charge, and containing almost all of the mass, surrounded by very light electrons occupying energy shells. But again there were problems. If the electrons were stationary, electromagnetic attraction would pull them into the nucleus. If the electrons were circling the nucleus at a sufficient velocity, this fate could be avoided, but another problem was presented: according to Maxwell, any charge that is accelerating will emit electromagnetic radiation and lose kinetic energy, which means that the electron would still spiral into the nucleus. At this point, Bohr and quantum theory entered the picture.

Bohr made two assumptions. The first was that a certain minimum orbit exists for the electron. Second, an electron in a higher orbit will jump to a lower orbit that is not fully occupied. In doing so, it radiates the full energy difference between the two orbits.

This led to the postulation that only certain frequencies could be emitted and implied a discreteness in all forms of energy. For various reasons, it became more and more difficult to picture electron balls orbiting around the nucleus, so that conception gave way to the idea of "energy levels," with electrons jumping from one level to the other and emitting or absorbing quanta of the proper frequency. The Bohr atom was an essential halfway step between the mechanistic physics of the nineteenth century and the new physics to come in the 1920s, yet it failed to explain why electrons occupy given energy levels or why they jump from one level to the other. Bohr's quantum jumps, as Wolfgang Pauli noted, were new wine put in old wineskins.

Meanwhile, an idea appeared that was to again revolutionize physics. Louis de Broglie began thinking that the photon had mass. While that thought has since been discarded, it led de Broglie to a discovery that changed physical concepts in a major way. Since, according to relativity theory, mass is energy, and Planck's and Einstein's conception of the quantum shows that energy is related to frequency, particles, de Broglie reasoned, are edging suspiciously close to being like photons. And since photons can also be waves, then electrons too can act as waves. From these ideas, de Broglie developed an equation stating the relationship between the wavelength of a particle and its mass.

De Broglie's equation can be applied to any moving body, whether an electron or a planet. But the relationship between wavelength and mass is mediated by Planck's constant which, as we have seen, is quite small, and ordinary bodies have wavelengths that are too small to be detected. However, on the electron level, the mass is small enough that the wavelength is similar to that of an x-ray. (It is important to emphasize that matter waves such as we are discussing are not electromagnetic in nature, even though they can exhibit wavelike properties.) In 1927 Clinton J. Davisson and Lester H. Germer used electrons to create diffraction patterns similar to those of x-rays, and this was considered proof that, as de Broglie suggested, an electron could behave as a wave. Einstein became aware of de Broglie's discovery and told Erwin Schrödinger about it.

Schrödinger envisioned the electron as a standing waveform circling the nucleus of the atom and developed an equation to describe this. The equation became known as the Schrödinger equation and its solution is known as the wave function. Only those orbits that could accommodate whole wavelengths were considered, and each energy level represented a different standing wave. The ground state housed only one wavelength, and no lower level of energy was possible. This all seemed quite plausible, and when Schrödinger applied his equation to the hydrogen atom, it worked well. But the hydrogen atom has only one electron outside the nucleus. When the same technique was applied to the helium atom, which has two electrons, the interpretation became difficult.

Max Born found that the equation describing the helium atom had to be written in a six-dimensional space. We know from classical physics that any object is located by three numbers in a Euclidean space. If two electrons at different points in space are mutually interactive, the equation for both electrons treats them as if they were one system occupying a given state, and six numbers are needed to describe this state. In this way, the system of the combined electrons becomes a wave in a six-dimensional continuum.

The concept of multidimensional space will be discussed in more detail later, but for the physicist of the 1920s, adding three dimensions for each additional electron had astonishing results. When the uranium atom is considered, with its atomic number of 92, the dimensions go up to rather large numbers. Even more mindboggling is to consider the wave function for the whole universe: the number of dimensions needed would be three times the total number of particles in the universe!

De Broglie envisioned the wave as a guide to the particle,

with both grounded in space and time. Schrödinger's approach, on the other hand, removed the electron to some sort of fictional multidimensional space and disentangled the wave from the particle. This in turn led Born to believe that Schrödinger's waves were not real — that is, not three-dimensional waves as Schrödinger initially thought. Instead, Born proposed that they were probability waves which rose and fell, giving the probability of a particle being in a particular place. But this probability wave flowed in a continuum of many dimensions. In addition, this probability system may be very different from what is intuitively assumed.

For example, let's say that within a given community, the population is 60% Protestant, 20% Catholic, and 20% other religions — that is, if you asked each individual in the community his or her religion, for every 100 people, the distribution would be as above. In other words, statistical knowledge comes from adding up our knowledge of individual cases. This is not true in the world of quantum theory, where statistical laws are all we have. It is impossible to describe the position and momentum of individual particles and then predict their future. (This will be discussed more fully in the next few pages.) Quantum laws deal largely with groups of electrons, not with individuals.

In the analogy above, it would be as if each member of the community could change religion every time he or she was asked, but probabilities for the entire community would always break down into the same percentages. Similarly, an individual electron has "freedom of choice" as long as the group follows statistical laws. (How the electron knows that a choice will fit in with the group law is considered later in the text.) French physicist Jean E. Charon uses an interesting metaphor to describe the historical journey that brought physics from classical certainty to quantum uncertainty:

> Science is like a visitor who looks at a crowd on Earth from a very high altitude. To study the crowd's behavior, this visitor will begin by studying its overall movements, the most visible ones and also the ones that are repeated; and he will derive from that, no doubt, great statistical laws relative to its behavior. Then, as he analyzes the crowd's behavior more deeply, he will at some point be obliged to study the behavior of the individuals who compose this crowd. And then, of course, he will no longer find statistical phenomena, and he will also see non-repeating phenomena. (Charon, pp. 132-133)

In Bohr's model of the atom, the electron jumps from one orbit to another by emitting or absorbing a quantum of energy of the proper frequency. In Schrödinger's approach, if no energy is entering or leaving the atom, the wave continues to propagate around the nucleus at a mean radius determined by the energy level of the particular electron. If the atom gains energy from the outside, the wave gradually flows into a higher orbit. During this process the particle can be found on one level or another at any given moment, while the probability of one level versus the other changes. That is, the electron wave moves continuously between quantum states. This flow of the wave function is not through space-time but rather through an underlying hidden domain. We are forbidden, by Bohr, to picture any detailed motion of the electron during the quantum jump. In this sense, the quantum jump is discontinuous. This discontinuity is the first fundamental way in which quantum theory is incompatible with relativity theory.

The second difference between quantum theory and relativity theory is that relativity requires strict determinism whereas quantum theory discards it.[15] Determinism implies that the future

[15]Strictly speaking, the Schrödinger equation is deterministic. Probabilities arise when the quantum world is elevated to the macro world. This will be discussed at length later in the text.

can be predicted, which requires that the present be knowable in all of its aspects. Schrödinger's probability wave (as discovered by Born) makes such knowledge unlikely, but the indeterminism of quantum theory is laid bare with the introduction of Heisenberg's uncertainty principle.

Basically, the uncertainty principle states that if you measure the position of a particle within a certain level of accuracy and *simultaneously* measure its momentum, the product of the two errors in accuracy cannot be smaller than Planck's constant. This can be visualized in the following manner. The position and velocity of an electron can be measured by shining light on the particle. The scattering of the wave of light indicates the position. But the accuracy of the position is dependent on the wavelength — the shorter the wavelength, the more accurate the measurement. Now, recall Planck's concept that energy comes in discrete quanta and is proportional to the frequency. As the wavelength decreases, the frequency increases and so does the energy of the quantum. Therefore, as you zero in on the position of the electron by using a higher-energy quantum, you disturb the velocity.

Thus, Heisenberg's uncertainty principle states that knowledge of both the exact position and the exact momentum required to predict the path of a particle (the determinism of classical theory) is not possible in quantum theory. The same indeterminancy can be exhibited for energy and time. It is tempting to assume that an electron has a definite position and a definite momentum even though, for various reasons, we cannot measure them with complete precision. But this is not the case. This lack of determinism is inherent in the structure of matter itself and cannot be blamed on the limitations of physicists' instruments.

The uncertainty principle leads to the idea that one electron (or any particle) is in principle indistinguishable from any other. It is not possible, because of the uncertainty principle, to precisely define the trajectory of each electron if two electrons are fired at each other. Remember that the position and momentum of an electron cannot be known precisely at the same time; we have only a probability distribution for both the position and momentum — as if the electron were "smeared" across a classical trajectory. Nor is it possible to tag an electron so that it can be tracked through a collision with a second electron. When electrons interact, the smearings overlap so that separate electrons cannot be identified, and there is no way of knowing if one electron is exchanged for another. It's as if a material body is made up of a *pattern* rather than discrete particles. Thus, matter loses its classical characteristics.

The third difference between relativity theory and quantum theory is that relativity is strictly local while quantum theory has a nonlocal attribute. Particles that are separated have long been known to influence each other by the forces of gravity and electromagnetism. With the advent of quantum theory, weak and strong nuclear forces were added to the list, but all these interactions are limited by the velocity of light. Less well known is an interaction not limited by the velocity of light, a nonlocal exchange: once two particles have interacted, they are, henceforth and forever, connected instantaneously by some sort of quantum bond. This bond, while originating as a mathematical construct, attained real status with John Stewart Bell's theorem, which states that after two particles have interacted, they continue to influence each other. This connection is not mediated by a field, but rather a portion of each particle is left with its partner and thereby seems to remain in constant contact, creating a kind of quantum wholism. This awesome and mysterious connection must be outside of space-time since the quantum bond is not consistent with the special theory of relativity and its restriction of the velocity of light. Thus, particles may be seen as projections from some sort of hyperspace, or underlying realm — the hidden domain of our title.

Einstein's formula $E=mc^2$ is the mathematical description of the equivalence of mass and energy. The amount of mass is a measure of the quantity of matter. Basically, the formula tells us that matter is frozen or locked-up energy; if this frozen condition is released, matter turns into energy. Conversely, if energy is restricted to a small enough volume, matter is created. This notion led some physicists, primarily David Bohm, to redefine the concept of light.[16] The underlying region described by the quantum wave function (called the implicate order by Bohm) is nonlocal, as indicated by the quantum bond. Light also is nonlocalized. That is, if someone could ride a light beam, it would appear that time stands still. This hidden domain can be seen as an ocean of light in which some light rays move back and forth rather than straight ahead, and it is this back and forth movement that forms particles. How this happens is not completely understood because physicists do not have a theory covering this underlying region. But in essence, this view indicates that matter is arrested or frozen light floating in a sea of light, like an iceberg floating in the ocean.

Finally, in quantum theory, a system is "questioned" by engaging it with a measuring instrument. The system then becomes part of a greater context which includes the apparatus. From a classical point of view, the system should not change its character when its context is broadened. In the quantum world, however, a system responds to a change of context in a manner similar to an organism. By definition, an organism is a complex system which not only functions in relationship to the properties of its parts but also responds to the character of the whole. Thus, according to quantum theory, the physical world exhibits some "lifelike" or organic characteristics.

---

[16]See *Bridging Science and Spirit*, pg. 86.

Based on this historical summary of quantum theory, the following points are most important for our subsequent discussion.

- All energy or action is in the form of discrete quanta (hence the name quantum theory). That is, all energy connections appear in small and indivisible units.
- The whole universe is interconnected by a network of quanta, like a sandy beach which from a distance looks continuous and indivisible but up close is seen to consist of minute grains of sand.
- Quantum theory is built on statistical laws based on probability, not on the laws of individual particles.
- The nature of all matter and energy is dualistic in that a system acts like a particle or like a wave, depending on the experiment.
- The hidden domain is nonlocal. The whole universe contributes to the organization of its parts; all constituents influence all others regardless of distance.
- At the most basic level, matter acts in an indeterminate way. The uncertainty principle indicates that the physical world exists in a world of potential, waiting to be activated by the physicist's equipment.

PART 2: THE DOMAIN OF THE WAVE FUNCTION

The mathematical framework of quantum theory has at its core the wave function, the most salient feature of which is that it does not exist in our three-dimensional world. Another way of saying this is that the wave function does not contain measurable energy and therefore is not accessible to our physical instruments.

That does not mean, however, that the wave function does not have measurable effects in our world. Evidence for its existence is provided by meter readings and computer readouts that conform to its mathematical description. Thus, it would seem that we live in a world of at least two levels: the wave function is on one level and correlates with objective effects in the other level, which is ordinary three-dimensional reality.

In the next ten chapters, we investigate the key concepts necessary for a thorough understanding of the wave function. The concluding chapter summarizes these ideas and discusses their philosophical implications.

3 The wave-particle duality

Does light consist of waves or particles? This problem has been with classical physics since its beginning. Newton said light was made from particles. The Dutch scientist Christiaan Huygens (1629-1695) said light was made from waves. Which is it?

Eighteenth-century physicists largely adopted the particle theory, the view favored by Newton. The tide turned against Newton in 1801, however, when Thomas Young performed the famous two-slit experiment, which we will discuss shortly. For now, let's simply say that the Young experiment established the wave nature of light. Nineteenth-century physicists adopted the concept of an all-pervading ether to carry the wave, but with the arrival of Einstein's theory of relativity it was discarded — or at least renamed. The vacuum now became known as space-time.

As mentioned earlier, in 1900 Planck came up with a formula that made it possible to solve all the problems encountered in the study of black-body radiation. He found that energy must be radiated in discrete quantities which he called quanta. Energy came in quanta only when entering or leaving matter; outside of matter energy was still considered a wave. That is, according to Planck, matter was required to absorb or give off energy in quanta.

Also, remember that while working on the photoelectric effect, Einstein found that light behaved as a quantum even after it left matter; the quantum of light was subsequently called a photon. So, the battle between wave and particle ended up with a truce. That is, depending on the experiment, light could be a particle (photoelectric effect) *or* a wave (Young's experiment). Later, de Broglie complicated the issue by reasoning that since light is a form of energy and, according to Einstein, matter is equivalent to energy, then if light could sometimes be a wave, so could matter. In this fashion, the theoretical existence of matter waves was established. When Davisson and Germer demonstrated in 1925 that electrons could exhibit wavelike properties, the confusion of the wave-particle duality engulfed matter as well.

With this review as background, let's turn to Young's famous two-slit experiment — the best known experiment for displaying the peculiarities of quantum theory, containing all the paradoxes associated with quantum mechanics. In Young's arrangement, a source of light was placed behind a small slit on a screen. The light beam passed through the slit to fall upon a second screen containing two slits side by side, through which the beam of light passed again. A third screen displayed the illumination that passed through the two slits. From each slit on the second screen came a diverging beam creating two areas of light, partially overlapping, on the third screen.

If light consists of particles, the region of overlap should be considerably brighter than the other areas, since it would receive particles from both slits. Surprisingly, that did not occur. Instead, the third screen showed alternating bright and dark bands. The only way this result could be explained was for the light from the slits to be waves. When waves are in phase, their crests match, and they reinforce each other; when waves are out of phase, a crest meets a trough and cancellation takes place. Thus, when the light waves from both slits were in phase, they reinforced each other and the illuminated region was bright; when the beams were out of phase, the result was a darker region. This pattern of light and dark bands, called an interference pattern, is a telltale sign of waves interfering with each other. In fact, by measuring the spacing of the bands, Young actually calculated the wavelength of the light.

Physicists today use Young's arrangement for a striking

display of the mystery of quantum physics, because it indicates how strange the wave-particle duality of a subatomic particle really is. Richard Feynman remarks:

> I will take just this one experiment, which has been designed to contain all of the mystery of quantum mechanics, to put you up against the paradoxes and mysteries and peculiarities of nature one hundred percent. Any other situation in quantum mechanics, it turns out, can always be explained by saying, "You remember the case of the experiment with the two holes? It's the same thing." I am going to tell you about the experiment with the two holes. It does contain the general mystery; I am avoiding nothing; I am baring nature in her most elegant and difficult form. (Feynman, *The Character of Physical Law*, p. 130)

To visualize subatomic particles as classical particles in the two-slit experiment, imagine that we are shooting marbles at the two slits instead of light. (Of course, the two slits will have to be large enough to allow the marbles to pass through.) If the detector screen records the pattern created by the marbles striking it, the results would be obvious and classical: two areas of hits slightly larger than the two slits, since the slits were slightly larger than the marbles.

Suppose then we use electrons (reducing the size of the two slits accordingly) with a photographic plate as the detector screen. Like the marbles, each electron that strikes the screen produces a confined and definite mark. But as we keep firing electrons, a strange thing happens: the pattern recorded begins to show wave interference. If we pause for a moment and reflect upon this situation, we are saying that when the electrons hit the photographic plate they were distinct particles, but before their detection at the photographic plate, they seemed to be waves.

It might have occurred to you that possibly the particle

electrons interacted in some way so that an interference pattern was produced, and that when we understand this interaction, we will see how the classical conceptions apply. Unfortunately, that is not the answer. If we fire the electrons slowly, one at a time, eliminating any possible interaction between them, an interference pattern *still* develops. If we duplicate this experimental arrangement throughout the world with many experimenters, each firing just one electron at a given prearranged time, with each individual photographic plate showing the arrival of the one electron, and the results from all the plates are added together, then, amazingly, the interference pattern shows up again! These experiments are arranged so that no signal can travel between them at less than the speed of light, so there can be no physical communication between the electrons. Just how does each electron know where to strike the plate so that an interference pattern appears? The answer to this question takes us into deep waters indeed.

It seems, strangely enough, that for such a pattern to appear, a single electron would have to travel through both slits and then interfere with itself. We can check that assumption by placing a recording device at each slit. Then we let loose with one electron from our source and record which slit it went through. If an electron is a particle, as its arrival at the photographic plate would indicate, we must assume the electron passed through one slit or the other, not both; a fraction of an electron has never been seen, so it cannot break up and go through both slits. We can show that the electron did not interfere with itself, but we pay a price for that knowledge, because by showing which slit each electron passed through, we have obliterated the interference pattern. That is, by looking at the electron, its wave properties are destroyed. Thus we cannot escape the conclusion that *if we do not peek at the electron, it acts like a wave, but as soon as we look, it becomes a particle*. In short, a measured electron is always a particle, but an unmeasured electron is a wave.[1]

Keep in mind that the unmeasured electron is a quantum wave, and that these are not waves that travel in space-time but are waves of potential. If this seems peculiar and counterintuitive to you, you are in good company. We have read what Feynman said about the two-slit experiment. This is what he said about quantum mechanics:

> There was a time when the newspapers said that only twelve men understood the theory of relativity. I do not believe there ever was such a time. There might have been a time when only one man did, because he was the only guy who caught on, before he wrote his paper. But after people read the paper a lot of people understood the theory of relativity in some way or other, certainly more than twelve. On the other hand, I think I can safely say that nobody understands quantum mechanics. (Feynman, *The Character of Physical Law*, p. 129)

If you think that Feynman is exaggerating, consider the anecdote Steven Weinberg relates in his book *Dreams of a Final Theory*:

> A year or so ago, while Philip Candelas (of the physics department at Texas) and I were waiting for an elevator, our conversation turned to a young theorist who had been quite promising as a graduate student and who had then dropped out of sight. I asked Phil what had interfered with the ex-student's research. Phil shook his head sadly and said, "He tried to understand quantum mechanics." (Weinberg, p. 84)

---

[1]Bohr attempted to understand this wave-particle duality by developing a concept he called the principle of complementarity. Bohr did not see this situation as paradoxical since the wave and particle aspects of an elementary particle do not appear in any given experiment in a contradictory way. Both are seen as complementary aspects of one overall reality; which face nature shows depends on the question we ask. So if one were to ask what an electron really is, Bohr would say that is a meaningless question. More precisely, physics at present cannot answer the question.

We have a reality described by waves, but we can see its effects only by the pattern made on our measuring instruments by particles. This is a difficulty that cannot be circumvented since it is basic to quantum interpretation. Just what are these waves? That is the central question we face in this book.

The physicists who developed quantum theory were aware that they were shaking the philosophical foundations of their science. An explanation for this strange wave-particle duality, called the Copenhagen interpretation, was presented by Bohr in 1927 at the fifth Solvay conference in Brussels. The simplest statement of Bohr's view is this: *the quantum world is not real*. Bohr recognized that the quantum world is completely different from our normal everyday world governed by the familiar laws of classical physics. Though we have mathematical formalisms to describe it, the unreal world of the quantum relates to the real world only by an act of measurement.[2] In other words, quantum theory says that the quantum world is detectable solely through meter readings and computer readouts and can be interpreted only through the esoteric mathematical formalisms of quantum mechanics. Essentially, the Copenhagen interpretation says that the physical properties of a quantum system depend on *how* they are observed. This view was nothing less than a major revolution in physics.

Newton understood that in classical mechanics some sort of probe was necessary to measure a physical system. The probe might be quite subtle, as when we determine the location of a body merely by looking at it. But even here light must impinge on the object and be reflected to the eye, which is to say that to measure requires some

[2]Actually, Bohr said that before the physicist can understand the activities of an electron, he or she must specify the whole experimental context. That means the reality of the world of the electron is ultimately connected with the arrangement of the macro world. So the electron does not have meaning except in its relationship to the whole. This idea resonates with many mystical religious concepts such as those found in Buddhism and Taoism and is confirmed by Seth, as we shall see.

sort of interaction between the observer and the observed. This is codified by Newton's third law, which says when object A applies a force on object B, B applies the same force on A. So, to measure a physical system involves some sort of disturbance. In many cases — for example, light bouncing off the moon — the disturbance is negligible. When the disturbance is significant, the classical physicist assumed that it could be accurately quantified and thereby taken into account in drawing conclusions.

The Copenhagen interpretation overturns this classical view. Bohr clearly states that at the quantum level a particle does not have specific properties unless it is being observed with a measuring apparatus. There is no objective reality underlying our everyday world except when it is being observed. As incredible as that notion sounds, it is important to keep in mind that this is not a concept held only by some small mystical group of physicists: Bohr's Copenhagen interpretation is the explanation accepted by the majority of physicists for the mindboggling phenomena of the quantum world.

The mathematical basis of quantum theory arises from the conclusion that since we do not know what state a system is in before measurement, it must be in a superposition of all possible states. If we wish to understand the world of the quantum particle, we will certainly want to know about its characteristics as they are revealed by our measuring instruments — but that will not be enough. We must also try to understand what a particle is doing in between measurements, that is, in its wave state. In describing the two-slit experiment, Heisenberg noted the difficulty involved:

> . . . The concept of the probability function does not allow a description of what happens between two observations. Any attempt to find such a description would lead to contradictions; this must mean that the term "happens" is restricted to the observation.

> Now, this is a very strange result, since it seems to indicate that the observation plays a decisive role in the event and that the reality varies, depending upon whether we observe it or not. (Heisenberg, p. 52)

Since experimental data do not help us here, we turn to the mathematical formalisms. The mathematics is complex and esoteric, but perhaps we can glean some clues from it nevertheless.

In classical physics, mathematics is used to represent the attributes of a system. The formulae are taken at face value and assumed to be actual descriptions of the evolution of the system. In quantum theory, the situation is quite different. Here the mathematics is an algorithm for calculating the results of experiments, at least as far as the Copenhagen interpretation is concerned. Actuality is no longer considered, but has evaporated into the mists of the mystical. The main reason for this is that if the mathematics is accepted as accurately describing what *is*, the results are most bizarre. So Bohr decided that physics no longer would tell us how nature is, only what we can say about it. That is, modern physics at its very basis is the procedure of providing mathematical algorithms which correlate the results of experimental observations.

But suppose we dare take quantum theory at face value — what can we then say about the reality of this world?

If the electron is a wave when it is unmeasured, we must examine qualitatively the mathematical formalism that corresponds to the wave aspect. The basic equation of quantum theory is Schrödinger's equation, a differential equation for which the solution is the wave function. The wave function describes a quantum system. That is, it contains all possible information which relates to the system being examined.

The wave function differs from a classical wave in a number of ways. Classical waves are associated with a large number of particles, as for example, photons in electromagnetic waves. But the waves of quantum mechanics can describe either a collection of particles or a single particle. The fact that a single particle can be in two places at once brings us to a second equation, which is a probability function. By using the two equations, the physicist can calculate the average value of an attribute of the particle if the experiment or measurement is repeated many times. As far as experimental results are concerned, these equations are extremely successful.

That is all well and good, but how can we picture such an unmeasured particle? Bohr summed it up in his Copenhagen interpretation by saying that the properties of matter (whatever it is) on the quantum level do not exist separately in a given particle in a precisely defined form. Rather they are incompletely defined potentialities which become definite only in interaction with a classical measuring apparatus. The wave function then describes these potentialities and provides a procedure for calculating the probability of a given potentiality actually occurring. The probability does not refer to the chance that a given attribute, such as position, actually exists before measurement but to the chance that a certain position *will develop* when the particle comes into an interaction with measuring equipment. Heisenberg describes this as a world of potentia, a term borrowed from Aristotle's philosophy, meaning a tendency — something between the idea of an event and its actualization.

Let us now suppose that the wave function with its alternative possibilities does, in fact, give us a picture of reality. How might that look? As we mentioned, such a picture is likely to be so strange that most physicists prefer to see quantum theory (and the wave function) as a calculational procedure only and not as an objective picture of the world. That is, most physicists view the wave function as a symbolic device that can be used to make predictions about what they will "see" if certain conditions are specified, but they do not consider it a description of qualities that can be located in space-time. Therefore, the wave function is sometimes thought of as an incomplete description of "reality" in space-time. Space-time, on the other hand, can be interpreted as a scheme that allows us to order our sense experiences; as Einstein himself stated, "Time and space are modes by which we think and not conditions in which we live" (Wheeler and Zurek, p. vi).

To label the wave function as merely a symbolic device because it does not describe actual events in space-time is somewhat paradoxical. While it is true that conceiving of reality in terms of a space-time continuum (as in classical physics and relativity theory) proved quite useful in understanding macroscopic phenomena, it failed on the microscopic level. Thus, it is logical to consider space-time also a symbolic device. But if we take both space-time *and* the wave function as descriptions of reality, then that reality is a great deal richer than the classical physicist could possibly have imagined. One is reminded of the parable concerning the blind men and the elephant. The first man touches the leg and says the elephant is a tree stump. The second touches the tail and says it is a rope. The third touches the ear and pronounces it to be a fan. The world of the wave function and that of space-time are actually different parts of the same elephant, that is, different aspects of reality. Physicists are simply more familiar with, say, the feel of the leg of space-time than they are with the strange ear of the world of probabilities.

In his splendid book *Quantum Reality*, Herbert imagines an imaginary quantum physicist Max deciding to face the quantum facts. To do so, Max designs a chamber which is capable of taking him into the world of the quantum. Here is Herbert's description of Max's entrance into that bizarre world:

> Max suddenly drops through the world's phenomenal surface into deep quantum reality. Holy Heisenberg! Centuries of Newtonian certainties vanish in an instant. Solid objects melt into the undivided wholeness as he

> enters the Place Without Separation. Max mixes with the mystery when his subject/object membrane dissolves. In tune with totality, Max creates a new universe faster than light wherever he turns his omnipotent gaze.
>
> What's it like down there? Max's sister Maxine says it feels just like Schrödinger's equation, only more so. You've got to see it to believe it. Behind the high-security fences of Max's quantum lab, consciousness creates reality, quantum logic is spoken exclusively, and for the trip home you have your choice of a billion different universes. (Herbert, *Quantum Reality*, p. 56)

Herbert's whimsical yet serious description of the quantum world gives us an intriguing glimpse into that other aspect of reality where probabilities, rather than certainties, are the rule.

Here are the important points to remember about the wave-particle duality:

- ☐ Young's famous two-slit experiment highlights the bizarre nature of the quantum particle. At the detector screen, the electron is entirely particle-like in behavior; no "parts" of particles are registered. Yet the pattern on the screen indicates that when we are not looking at the electron (i.e., when it is passing through the two slits) it acts like a wave. We conclude that an observation *creates* a particle from the quantum entity which otherwise is a wave.
- ☐ According to the Copenhagen interpretation, the electron does not have specific properties until it is observed. That is, until we "look," there is no objective reality, but rather an underlying state of potential. If the wave function is an accurate description of this

hidden domain, then this potential is really a superposition of all possible states.

- ☐ It is our thesis that the wave function is not merely a symbol to be used in a calculational procedure. Rather, it has real meaning and describes a hidden domain that is the creative source of our three-dimensional world.

4 The significance of complex numbers

Before examining the mathematical formalisms of quantum theory, let us be aware that we are discussing *interpretations*; we are not questioning the accuracy or suitability of quantum physics as a theory. We shall start with simple descriptions of the mathematics involved and see where these lead us philosophically. That is, we will penetrate domains that are not accessible to physical instruments; but like Plato's ideal world,[1] these realms can be reached through the use of the mind. To enter these deep and mysterious regions, it is advisable to open not only our minds but also our hearts. Then, perhaps, to our amazement, we may discover a world more congenial, more beautiful, and more exciting than any we have ever known.

The first mathematical concept we need to understand is the complex function. In mathematics, the term *complex* does not mean "complicated." Rather, a complex number is defined as a number consisting of a combination of "real" and "imaginary" numbers — clearly a challenging notion. The fact that the wave function is a complex function is one of the basic reasons for the many peculiarities in quantum theory.

Real numbers include 1) rational numbers, consisting of the natural numbers 1, 2, 3..., their negative counterparts −1, −2, −3..., 0, and rational fractions; and 2) irrational numbers, which are quotients expressed as infinite decimals (these will not concern us here). Real numbers can be envisioned as points on a line, with one point designated zero so that all points to the left of it are negative real numbers and all points to the right are positive real numbers. But

[1]In about 360 B.C., Plato envisioned an ideal world accessed only though the mind. Mathematical concepts like triangles and circles that are only approximated in the physical world inhabit the ideal world in a perfect form.

since points are theoretically dimensionless, between any two points or numbers, there exists a third number. Therefore, the real number line contains an infinite number of numbers. Real numbers are especially useful in physics because physics is grounded in measurement, and the physicist needs real numbers to designate the magnitude of such attributes as mass, space, and time for a wide range of physical phenomena. For the most part, real numbers and physical reality seem to agree quite well, although at some inconceivably small distances, length may no longer be definable.[2]

To comprehend imaginary numbers we need to use simple algebra. Since the square of any number, positive or negative, is always positive, there is no number that when squared will equal a negative number; that is, a negative number has no square root. Mathematicians (who understood this as long ago as the sixteenth century) were unwilling to declare such an operation meaningless. Instead they postulated a new number for the square root of minus one, which they called "imaginary" (as opposed to "real") and which was designated by the letter i. To take the square root of minus four, we can write $\sqrt{-4} = \sqrt{4 \times (-1)}$. If $i = \sqrt{-1}$, then $i^2 = -1$. Substituting this into the previous equation, we have $\sqrt{-4} = \sqrt{4i^2} = 2i$. So by employing i, we can take the square root of a negative real number.

To fully understand the wave function, which contains imaginary numbers, we also need to grasp the relationship of imaginary numbers to physical reality. As we delve into this subject, we find that imaginary numbers are very useful in understanding how the world operates. We can place imaginary numbers on a line, as we did with real numbers, with negative numbers to the left of zero and positive numbers to the right. On the imaginary line we will have such

[2]Some scientists think the space continuum may actually be grainy, which implies a universal minimum length, usually estimated to be 10^{-33} cm. Every gravitational wave in space has a zero-point energy and because of that, the lengths below 10^{-33} become undefinable.

numbers as $2i$ and $5i$ instead of 2 and 5 which are on the real line. We can add and subtract imaginary numbers using the same operations as for real numbers: $5i-2i=3i$, just as $5-2=3$. But a peculiar — some would say fortuitous — situation arises when we multiply imaginary numbers: $5i \times 2i=-10$, which is a real number. The simple act of multiplication moved us from the imaginary to the real.

The relationship between the real and imaginary domains of numbers is elegantly demonstrated by mathematicians through a geometric display. If we place the real line horizontally on a page and the imaginary line vertical to it, with the zero points coinciding, we create a two-dimensional space referred to as the complex plane (introduced by Jean Robert Argand in 1806) (Fig. 1). This representation allows operations and relations involving complex numbers to be given a geometric meaning. Imaginary numbers are located on the imaginary axis, of course, and real numbers on the real axis. But what about all the points in the plane that are not on either axis? These points are a combination of real and imaginary numbers, or complex numbers: in the figure, the complex number $5+3i$ has an imaginary part $3i$ and real part 5.

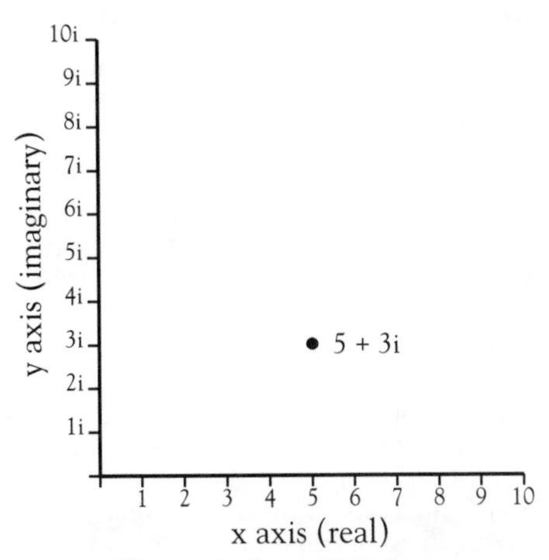

Figure 1. Complex plane

Seeing the real number line as embedded in a plane of complex numbers offers a tremendous enrichment of our understanding of real numbers, something like uncovering a whole new continent. For example, if a mathematical function contains a complex variable rather than a real variable, then the function also can be divided into real and imaginary parts. Thus, the complex number is a more complete concept since it encompasses the entire range of both real and imaginary numbers. If one starts with a complex number, a real number is easily produced by allowing the imaginary part to equal zero; and conversely, an imaginary number is one in which the real portion is zero. Thus, since the complex number contains all the imaginary and all the real numbers, its domain can be considered as primary. We shall return to this important idea later on.

It would be difficult to overstate the utility of complex numbers in physics. In quantum theory, complex numbers are absolutely necessary, as we shall see. Relativity theory also relies on complex numbers; for instance, some of the solutions to Einstein's field equations turn out to be complex. The role of i in designating the time coordinate in relativity was popularized in Stephen Hawking's best-selling book, *A Brief History of Time*. Hawking concludes that if one uses imaginary time rather than real time, the universe could be finite but does not have an initial event and thus is without a singular origin. This concept negates the need for a beginning and, in Hawking's view, eliminates the need for a Creator.[3]

The important point is that imaginary time is required in relativity theory, and this theory clearly eliminated the distinction

[3]A good way to envision the absence of a first event is to reduce time from the present back to the beginning point by halves to end up with zero time. However, as the Greek philosopher Zeno found, this is not logically possible. No matter how small the remaining interval of time is, it can always be halved again, never reaching zero. Also, due to quantum uncertainty or fuzziness, it is impossible to establish a first event. While we can speak of a boundary to time which is zero, that is not establishing an actual initial event. However, all this can be explained by accepting time as discontinuous rather than continuous.

between space and time. As you know, in Newtonian physics, space and time were considered to be separate entities; time was independent of space and its three coordinates. But relativity theory made it necessary to abandon the absolute character of time; the idea of objects in space with a separate, flowing time was replaced by the concept of events in the four-dimensional space-time described by Minkowski. Since any appearance of simultaneity is purely relative, physical reality can no longer be seen in terms of a point in space or an instant in time — only the four-dimensional event is real.

With time linked to space in this way, the distance between two points in ordinary Euclidean geometry becomes the interval between two events. The interval combines space and time in such a way that when motion makes lengths contract, the passage of time is slower, or, time expands. So, although two observers traveling at a high velocity relative to each other would differ on the spatial and temporal separation of two events, when they calculate the *interval* between the events, according to their own space and time measurements, the results are always the same. This tells us that the properties of space and time are in a sense merged and cannot be separated. That is, space by itself and time by itself are peculiar to each observer, but space-time is common to all. It is generally thought that it is impossible to break the velocity-of-light barrier. If matter could travel at superluminal speeds, then space would twist into time and time into space, and material objects could travel into the past.

To help us understand the role of imaginary numbers in relativity theory, let us recall the Pythagorean theorem, which defines the relationship between the lengths of the sides in a right triangle: the square of the hypotenuse (the side opposite the right angle) is equal to the sum of the squares of the other two sides. In a two-dimensional space defined by axes x and y, any point on the plane can be used to form a right triangle. If the point is S, the hypotenuse would be a line from the origin (the junction of x and y) to S, with one side of the right angle formed by one axis and the other by a line drawn perpendicular to it, so that $S^2=x^2+y^2$. In a three-dimensional space, $S^2=x^2+y^2+z^2$, if z is the third coordinate.

As we have seen, space and time are so inextricably mingled that time can be considered a fourth dimension. But to designate the time axis we must use a distance nomenclature, and in multiplying time (t) by the speed of light (c), we end up with a minus sign, $S^2=x^2+y^2+z^2-c^2t^2$.

In Minkowski's procedure for computing the space-time interval, although as we can see from $x^2+y^2+z^2$, distances are normally added, time had to be subtracted, thus leading to the concept of imaginary time. So, if a represents the fourth dimension, then $a^2=-c^2t^2$, and extracting the square root of both sides gives us $a=ict$; a is like the dimensions x, y, and z except it is imaginary. Now, $S^2=x^2+y^2+z^2+a^2$ holds true for all observers traveling relative to each other at a uniform velocity. To sum up, time is a fourth dimension but is designated by imaginary numbers rather than real ones.

As pointed out earlier, the distance of 300,000 km (speed of light per second) is a manifestation of time equal to one second. Some physicists, such as Wolf (quoted on p. 29, n. 8), see us marching on "through an imaginary dimension of space called time, moving at the speed of light." We can conjecture that there are two types of time. One type describes a sequence of actual events brought about by the collapsing wave function, which is the time we are aware of in the three-dimensional universe. The other kind of time, used to create space-time in Einstein's relativity theory, refers to the evolution of possibilities in the Schrödinger equation. Since possibilities are not real events, that type has been called imaginary or virtual time. Hawking puts it this way:

> A space-time in which events have imaginary values of the time coordinate is said to be Euclidean, after the ancient Greek Euclid, who founded the study of geometry of two-dimensional surfaces. What we now call Euclidean space-time is very similar except that it has four dimensions instead of two. In Euclidean space-time there is no difference between the time direction and directions in space. On the other hand, in real space-time, in which events are labeled by ordinary, real values of the time coordinate, it is easy to tell the difference . . . .
> (Hawking, p. 134)

To summarize, Einstein and other physicists found imaginary numbers to be extremely useful in describing space and time. That is, space and time are physical dimensions somewhat analogous to latitude and longitude but with the significant difference that one of the dimensions is measured along a scale of imaginary numbers. That means that if the mathematics of relativity expresses our universe accurately, one of our dimensions is imaginary, and we all are in some sense moving at the velocity of light. (Could it be that there are other universes with other values for the velocity of light? We shall consider that intriguing question later, in Chapter 15).

Since Schrödinger's equation is a complex function,[4] how can we obtain real numbers in our measuring instruments and calculations? It is not useful to know that two electrons are 100 complex centimeters apart. Physicists use a simple procedure to go from complex to real numbers, based on considering the wave function as a

---

[4]We might ask if it is necessary for the wave equation to be complex. After all, it is not too different from Maxwell's classical equations. If we can use a real equation for light waves, why do we need a complex function for electrons? First of all, there are also classical equations that are solved with complex functions, but the real and imaginary parts are independent of each other, and the complex function is just an auxiliary technique. In Schrödinger's equation, the real and imaginary parts are coupled, and neither of them alone is a solution to the wave equation. More important, both the real and imaginary parts contribute to the physical results obtained when computing probabilities. So, the wave function must be complex.

probability field. Before we discuss this procedure, we need to understand probability waves as used in quantum theory.

One of the first things we learn about quantum theory is that it is probabilistic. But Schrödinger's wave equation provides a *continuous* and *causal* prediction of how the wave function evolves in time, and thus it is completely deterministic. If the wave function describes quantum reality deterministically, what is the philosophical problem that made Einstein believe quantum theory is incomplete? Though Schrödinger's equation can be used to determine subsequent states of a wave function, the wave function is a *collection* of subsequent potential states, all of which evolve together deterministically in time. Only one state, however, can occur in our three-dimensional world. That is, an electron cannot be at points A and B at the same time, just as you cannot join a friend for lunch at a restaurant between noon and one o'clock and also be at home during that hour to hear your favorite radio program.

A second equation in quantum theory, taken together with Schrödinger's equation, provides the complete procedure for determining where the electron is. But this is a probabilistic equation that can be used to determine the likelihood of the electron being at point A, B, or C only if *measured.* Using these two equations together has been enormously successful, but it is significant that the only connection between them is an act of observation. In other words, if we do not observe the system, the wave function continues as probabilities, and the system remains a collection of states, with no particular state elevated to the level of three-dimensional reality; in the nonobserved world described by the wave function, an electron must be spread out rather than concentrated at a particular point. This idea made Einstein uneasy and causes many physicists to find quantum theory lacking since it does not provide them with a satisfactory and objective conception of the quantum world.

Normally, we talk about probabilities in percentages: we say there is a 60% chance of A happening and a 40% chance of B happening, or several alternatives with probability percentages adding up to 100%. In quantum theory, the probability percentages are complex numbers. If event A is said to have a certain complex percentage of happening, what does that mean? How are we to picture that? To get around this problem, the physicist does not call this complex number a probability but a probability *amplitude* — and amplitude is a property we associate with waves. If we assume that the unseen quantum system is a simple electron, then the electron is described by a wave that is not in space-time.

In 1798 Joseph Fourier discovered that any wave, no matter how complicated, can be written as a sum of unique and individual sine waves. Each sine wave can have its own frequency, amplitude, and phase, but added together, they are equal to the original wave.[5] All the waves in a Fourier series of the wave function are quite wavelike in the ordinary sense, except that they do not carry energy. With normal waves, the amplitude squared gives the energy of the wave. But what does amplitude squared mean for a wave that doesn't have energy?

When Schrödinger introduced his equation, he interpreted the wave function as a measure of how thickly the electron was spread out. This view was quickly seen as untenable, however, because whenever an electron is encountered, it is always whole and located at a specific position. In 1926, Born suggested that the wave function is a measure of the probability of the electron being in a particular place, resulting in the idea that Schrödinger's wave is a wave of probability. That is still the accepted view in the physics

---

[5]Today, with the aid of computers, physicists can disassemble a wave into other waveform families, not just sine waves; but for simplicity's sake, we shall discuss only sine waves.

community. Thus, the amplitude squared is no longer a measure of the energy of an electron but a probability of the electron being in a certain place.

If the waves in the Fourier spectrum are each to play the part of a probability, a problem arises. Since the amplitude is a complex number, if we square it to get the probability, we get another complex number. These complex amplitudes, then, are not actual probabilities but can be seen as the quantum mechanical analog of probabilities, called, as we said, probability amplitudes. To get a *real* probability number, we must turn to the formal procedure for creating real numbers from complex numbers, which we shall discuss at some length since it seems to point the way from the mysterious world of the wave function to our three-dimensional reality.

Every complex number has what may be termed a mirror image called a complex conjugate. Recall that in the notation for a complex number, $a+bi$, a and b are real numbers and the imaginary portion is bi. If the amplitude of a wave is $a+bi$, and we multiply it by itself, we get another complex number. But if we multiply $a+bi$ by $a-bi$, the result is a real number. In this approach, $a-bi$ is the complex conjugate of $a+bi$. So, to obtain the real probability for each term in a Fourier series, each amplitude is multiplied by its complex conjugate.

Let us pursue the idea that the world of the wave function permeates our real world and is described by complex functions, and that by multiplying such numbers by their complex conjugates we actually create the probabilities we associate with reality. In 1945 Wheeler and Feynman developed what came to be known as the absorber theory of radiation, which had to do with electromagnetic radiation, or light, not with electron waves. This theory starts with the idea that an atom emits light which is absorbed by a second atom at some time in the future. The light emitted from the first atom is called the retarded wave and travels forward in time. Maxwell's four equations describing the integration of magnetic and electric fields have two solutions: the retarded wave, which goes forward, and the advanced wave, which describes radiation going backward in time. Most physicists ignore the second solution. But Wheeler and Feynman assumed that the second atom, encouraged by the retarded wave, sends an advanced wave back in time to the original emitting atom. The result is that the two waves come together in such a way as to duplicate the wave of conventional radiation. Because the two waves mesh so perfectly, the advanced wave is never observed.

John Cramer published a series of papers in the 1980s that effectively incorporated the Wheeler-Feynman approach in describing quantum theory. Cramer reasoned that since quantum theory says the electron is also a wave, perhaps the approach suggested by Wheeler and Feynman could be used for the emission and absorption of particles as well. The basic tenet of Cramer's transactional interpretation of quantum theory (as it is called) is that every quantum event involves a kind of "handshake" between the past and future, so that in some way, the future is affecting the past. For example, when we see light from a star that is ten light-years away, the retarded wave has been traveling through space (from our perspective) for ten years. But when we observe the light, we reach into the past by ten years to complete the process.

Fred Alan Wolf in his book *Star Wave* relates Cramer's interpretation to the Schrödinger wave equation and the wave function. Wolf points out that the wave equation describes the evolution of the wave function with time, which requires establishing some initial time or boundary condition. The complex conjugate wave also is a solution to the wave equation, but its initial time or boundary condition is some time in the future. The wave equation also describes the evolution of the complex conjugate wave, which is going backward in time. You can see the similarity of this approach to the Wheeler-Feynman theory. In essence, we need both waves to create reality: one wave travels forward in time, and the other travels backward.

So, if we have two events A and B, and we assume that A causes B, we are assuming B occurs at a later time than A. But Cramer's transactional interpretation explains the two events as a cooperative process between A and B — the handshake occurring across time. As a practical example, assume that A is a light sending out a light wave and B is the light wave impinging on your eye. The light going from the source to your eye is the retarded wave going forward in time. However, you send an advanced wave back in time to complete the transaction. In a way, you agree to having the light hit your eye, and thus in some sense create event A. A human being is not necessary to create an advanced wave; an electron does the same thing when it "agrees" to jump from one orbit to another.

In this examination of a rather difficult mathematical concept, complex numbers, and its relation to the wave function, the following points are most important to understand:

- ☐ Complex numbers, used to describe the wave function, are an integral part of relativity and quantum theory.
- ☐ Since the wave function is complex, it is not observable in our everyday three-dimensional world. That is to say, an electron cannot be seen as a "real" standing wave.
- ☐ Although the wave function is not directly observable, it seems to underlie or permeate our real world.
- ☐ The complex plane apparently is not only a mathematical convention but refers to the very source from which the physical world is manifested.
- ☐ The product of the wave function and its complex conjugate may provide a clue regarding the process of physical creation, an idea we will pursue further in Chapter 14.

5 Multidimensionality and the wave function

The fundamental postulate of classical physics was that the world could be separated into basic units whose interactions could be analyzed. Essential to this approach was the conception of a void or empty space as the arena of activity for the basic particles. To place this notion on a geometrical footing, Descartes assumed the universe to be a three-dimensional container in which bodies move in relation to an evenly flowing time, in accordance with the dynamic laws of Galileo and Newton. The three dimensions are right/left, forward/backward, and up/down, represented in the Cartesian coordinate system by three axes perpendicular to each other. This system allows every point in space to be defined by an ordered sequence of three real numbers.

Newton's invention of the calculus made possible a cornucopia of technological achievements, and as the science of mechanics progressed, brilliant additions were made by others. One such contribution in the nineteenth century, by William R. Hamilton, led to an important concept called phase space. In Newtonian physics, if the position, velocity, and mass of a physical body are known at some time t_0, in principle, at some later time t_1, all three attributes can be computed. The position and velocity may change, of course, but the mass remains constant. The basic variable is position, and velocity is derived from position; that is, velocity is the rate of change of position relative to time. In the Hamiltonian scheme, however, position and velocity are seen as independent variables. Furthermore, Hamilton used the concept of momentum — the product of the mass of an object and its velocity — rather than velocity alone. Then he reformulated Newton's mechanics into two basic equations, one describing the momentum of a group of particles and the other their positions. These represent the total energy of the system in terms of the two variables, and this energy is expressed in a geometric notation called phase space.

If, as Hamilton postulated, position and momentum are both primary, in addition to the three coordinates required to define position, three more dimensions are required for momentum, so we have two sequences of three real numbers for a single body. If six dimensions are needed to describe one particle, then ten particles in a system would require sixty dimensions. That is, each point of phase space can represent the entire system, and as the point moves, it describes the evolution of the system with time. But the evolution takes place in a multidimensional domain. This cannot be visualized geometrically, but if you accept phase space as a convenient arrangement of information — which it has turned out to be — then visualization is not necessary.[1]

In quantum theory the analog of phase space is Hilbert space. As in Hamilton's concept, a single point in Hilbert space (albeit in different mathematical dress) represents the entire quantum state. But the differences between Hilbert space and phase space are important. Hilbert space is a multidimensional complex vector space. It was developed by David Hilbert in the early part of the twentieth century for purely mathematical reasons. In the 1930s, John von Neumann consolidated ideas from Bohr, Heisenberg, and Schrödinger and placed the new quantum theory in Hilbert space. Hilbert space, like phase space, can be infinitely dimensional and therefore can handle all the possibilities of a quantum system in one convenient package.

In Hilbert space, a vector represents the Schrödinger wave

[1]Space with dimensions greater than four (assuming one for time) is known as hyperspace. Scientists and philosophers have long been aware that such spaces are possible. A circle of two dimensions is a cross section of a sphere, a sphere is a cross section of a four-dimensional form, that form is a cross section of a five-dimensional form, and we can go on to infinity with this procedure. A circle is also a projection of a sphere in two dimensions. Likewise, we in our three-dimensional space may be seen as projections from a higher-dimensional space. In this fashion, we can describe God or All That Is (Sethian term) as dwelling in an infinite-dimensional space and ourselves as projections from that source.

function. A vector is a mathematical entity that exhibits both direction and magnitude. It is represented pictorially by a straight arrow whose length is proportional to the magnitude of the quantity in question and whose point indicates the direction. The point of the arrow and a point in geometrical space coincide, while the tail of the arrow lies at the origin of the coordinates. Unlike points, vectors can be added and multiplied, which is essential for quantum theory. Two vectors of the same length but pointing in different directions represent two different quantum states.

Figure 2, a two-dimensional vector space, will help us grasp some of the basic properties of vectors. The two axes are *x* and *y*, and the vector is represented by the line *op*. The letter *o* designates the origin of the two coordinates. From the point of the vector, a perpendicular line is drawn to each axis, resulting in lines *a* and *b* as projections of vector *op*. Each projection is now a vector in itself, located along each coordinate (Fig. 2A). Vectors *a* and *b* are added together by placing the tail of vector *b* at the point of vector *a* and the point of vector *b* at point *p*. The result is a right triangle where the legs are vectors *a* and *b* and the hypotenuse is vector *op* (Fig. 2B). Thus, vector *a* plus vector *b* equals vector *op*.

From this simple diagram (Fig. 2), we are going to make a crucial generalization. A single vector can be seen as the sum of any

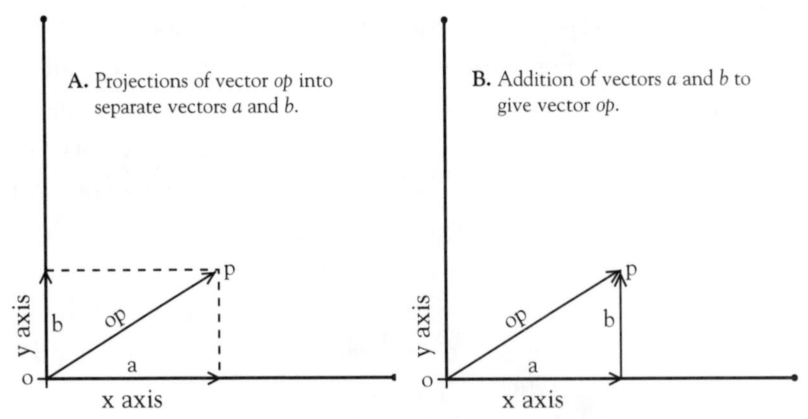

Figure 2. Two-dimensional vector space

number of vectors depending on the dimensions of the space we are using. All these vectors making up the single vector will be at right angles to each other. When we attempt to understand a vector in Hilbert space, we will use this simplified analogy. That is, a vector in Hilbert space (referred to as the state vector) can be expressed as the sum of the projected vectors on each axis. However, since Hilbert space is really a complex vector space, the vector (or ray, as it is sometimes called) is really a complex line. (To avoid complications, we shall not concern ourselves with that concept now.) Recall that a vector in Hilbert space characterizes a quantum mechanical state. So, we are saying that a quantum state is the same as a combination of other quantum states — like the Fourier analysis of a particular wave function, but using vectors instead of waves. Von Neumann's accomplishment was to make Schrödinger's wave function a vector in abstract Hilbert space; both express the same quantum state but in different mathematical terms.

It is evident that Hilbert space does not represent a void as in Newtonian mechanics; it is not merely adding dimensions to three-dimensional space. Rather, Hilbert space is a mathematical device for arranging pieces of information, with each coordinate representing a possibility for a given quantum state and corresponding to a distinct value for energy, position, momentum, spin, etc. Now, if the vector in this space is not parallel to any axis, it does not have a definite value for these attributes, so the idea that a vector can be decomposed into a series of vectors perpendicular to each other is crucial to quantum theory. Each of these perpendicular vectors represents a particular potential behavior of a quantum system. But since all the vectors are perpendicular to each other, none of them has a component along any others, which is to say that all potential activities are distinct. Also, the number of dimensions needed in this abstract space corresponds to the number of choices available for the quantum system, and this can go to infinity.

If we are talking about the quantum state of an electron,

then each axis is a possibility of finding an electron at a given position at a certain time *t*. As the vector representing the state moves and changes direction, the magnitudes of all the possibilities also change. That is, the projection of the state vector on each axis is a measure of the possibility of an electron being at a particular position. In the classical view, it was assumed that an electron at any given time would be a vector along one of the axes. That is to say, an electron cannot be in more than one place at a time. But in Hilbert space, that is not true: the electron is in all possible places all the time. Therefore, Hilbert space cannot be describing our familiar everyday world. Rather it is describing the world of the wave function which underlies and permeates our three-dimensional world.

So far, we have no means of selecting a possibility in order to create a real event. But in Hilbert space there is an analog for multiplying the wave function by its complex conjugate to obtain a real probability. Just as the wave function is represented by a vector in Hilbert space, so is its complex conjugate also in Hilbert space, but it turns out to be the mirror image of the original vector. To see this, let's place the original vector from Figure 2 in a complex plane, where *x* is a real axis and *y* is an imaginary axis (Fig. 3). In addition to vector *op* we now have vector *oq*. Vectors *op* and *oq* are equal in

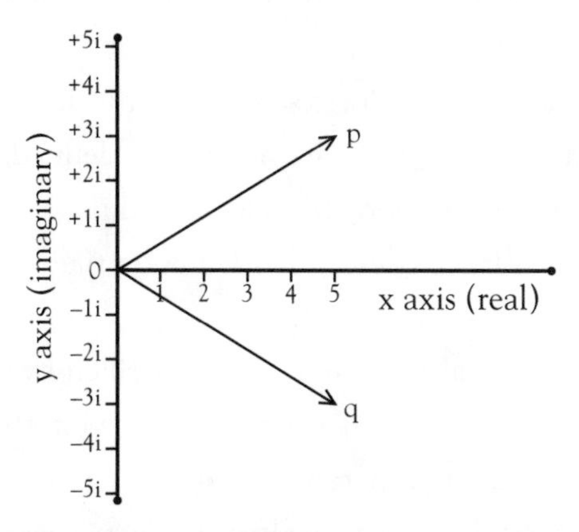

Figure 3. Multiplication of vector by its complex conjugate

length, and the angles they make with the real axis are equivalent, so *oq* is the complex conjugate of *op*. The real axis divides the complex plane in two, with everything above the real axis positive and everything below negative. Thus *oq* can be seen as the mirror image of *op*, with the real axis acting as the mirror.

Recall that mathematicians have a procedure for obtaining real numbers from complex numbers by multiplying the complex number by its complex conjugate. In a similar manner, multiplying a vector by its complex conjugate produces a real number. So, when a quantum state is measured, the quantum state "jumps" to one of the axes of Hilbert space determined by the measurement. The choice is constrained by the probability obtained when the various possible states are multiplied by the complex conjugate. Keep in mind that there is no law which informs us of the choice. Rather, the selection is random but is restricted to the various calculated probabilities. So, in some sense, when a vector and its complex conjugate are combined, a real event occurs.

Hilbert space contains infinite dimensions, but these are not geometric. Rather, each dimension represents a state of possible existence for a quantum system. All possible states coexist and add up to the wave function before a measurement is made or a given possibility is selected. The electron unmeasured is a complicated pattern in an infinite-dimensional Hilbert space; it is not in our three-dimensional space, and its attributes are not clearly defined. That is, in the world of the wave function, the electron does not have inherent properties but has incompletely defined potentialities.

To summarize:

- ☐ Originally, classical mechanics was described by Newton's three laws, and these were geometrically expressed in the three-dimensional Cartesian coordinate system.
- ☐ Hamilton formulated a new description of classical mechanics

which was eventually housed in an infinite-dimensional phase space. In this space, a point represents the entire physical system.

- ☐ Von Neumann set quantum theory in an infinite-dimensional complex vector space called Hilbert space. In this space, a vector represents a quantum state, as does the wave function in Schrödinger's formulation.
- ☐ Each dimension in Hilbert space represents a *possible* state for a quantum system, so an unmeasured electron exists as a very complicated pattern.
- ☐ A *particular* state is selected when, after the measurement, the state of the system "jumps" to one of the axes of Hilbert space based on the concept of the complex conjugate. There is no law that tells us which axis will be selected. Rather the choice is random, but the probabilities are strictly followed.
- ☐ The wave function description in Hilbert space cannot by itself tell us which state will be selected for reality. The action of the complex conjugate in obtaining real numbers offers a clue as to what else is needed to bring potentiality into actuality. This point is basic to the so-called measurement problem and to our understanding of what underlies our three-dimensional world.

6 The linear superposition of states

Normally waves require some sort of substance, such as water, to undergo vibrational motion. But even though the quantum wave is a vibration of *possibilities* rather than of a material substance, waves are waves and pretty much follow the same rules. One of the basic principles governing waves is the superposition principle, which states that the amplitude of a complex wave can be determined by adding the amplitudes of its component waves.

A rock thrown into a still lake pushes the water down where it hits, creating a depression, which causes the displaced water to rise around the point of contact. But then the raised water is also pulled down by gravity, and it too displaces water, so another ring of water is raised around the second depression. In this manner, the *disturbance* caused by a rock thrown into the water — not the water itself — travels across the lake. The water itself moves *only* up and down. This can easily be verified by placing a wood chip on the water as a wave comes through: the chip goes up and down, not across. If a second stone is thrown into the lake, it creates a second wave, which is superposed onto the first, resulting in a rather complex wave[1] compared with the original wave. But mathematically the amplitude of the complex wave is described by a simple rule: add the amplitudes of the two waves at every point on the lake, and the result is the amplitude of the complex wave. This is the principle of superposition. If the amplitude becomes too large, waves in a material substance do not obey the superposition principle and are said to be nonlinear. Quantum waves, however, always obey the superposition principle, and are never nonlinear when undisturbed. A linear

[1]In this case, we are using "complex" in its usual sense as intricate or complicated. Water waves do not create waves which are described by complex numbers.

system is one in which the whole is equal to the sum of its parts. Thus, linearity means that by understanding the parts of a system, one can, in principle, understand the whole.[2]

Now, suppose two waves are heading towards each other, with amplitudes in the range such that the superposition principle applies. When the waves meet, a complex wave is created according to the rule given above. If the waves then separate, the original amplitudes are retained (barring friction). That is, no distortion was introduced, and the wave is linear. Thus, a complex linear system can be decomposed without destroying the system. You can see that linear systems lend themselves to mathematical treatment because they can be reduced to a simple superposition of elements.

As we have noted, in classical mechanics, if the initial position and momentum of a system are known, then theoretically any subsequent state of the system can be determined. But in quantum mechanics, an initial system potentially can develop into any number of subsequent potential states. The ensemble of these potential states is represented by the wave function. All these states change with time in a deterministic fashion, according to Schrödinger's equation. There are no probabilities involved at this point, which means that the wave function is in a superposition of states and evolves in a linear fashion. Nothing in the Schrödinger equation tells us how to undo the superposition or make the wave nonlinear.

As we have seen, Bohr attempted to solve this problem by saying that an observation with a measuring instrument is the connection in that it changes the linear wave equation into some specific state, making it nonlinear and thus transforming the

[2]Because a linear system can be decomposed into its parts, the mathematics required is relatively simple. If physics dealt only with linear systems, analysis would certainly be easier. But nature is not constructed that way; real systems are often nonlinear, with the whole being greater than its parts. This very nonlinearity is a source of nature's richness. In fact, many physicists see in nonlinearity a creative and therefore lifelike quality, as opposed to Newton's mechanistic view.

quantum system to the material level. In effect, Bohr divided the world into two distinct realities: the first, described by the wave function, is a realm completely closed to our experience, while the second, described by interactions between particles, belongs to the classical realm and is the essence of our everyday lives. That is, the first reality consists of waves, which we cannot directly see but whose effects we experience as particles. Rather than talking about esoteric measuring instruments like bubble chambers,[3] let's take one instrument that we all possess — the human eye and its attached optical system. That is, our eye, as a measuring instrument, changes the waves of this first reality into a photon, which is the particle aspect of light.

Von Neumann, questioning Bohr's division of reality into two realms, wanted to find the exact point at which a potential state becomes actual. This occurrence is referred to by physicists as the collapse of the wave function. Breaking down the act of measurement into small steps, von Neumann reasoned that the world is quantum-mechanical throughout: if the measuring instrument is made of quantum particles, it too can be represented by a wave function. Where, then, does the collapse occur?

Von Neumann's model of the measurement process was something like this. If the initial quantum system is an electron, then the electron and the measuring instrument together become a larger quantum system satisfied by a kind of superwave function. While the wave function of the electron would certainly be collapsed, such would not be the case for the wave function of the whole. This need for an ever-larger context leads one finally to the brain, which also is

[3]The bubble chamber is an instrument that records the track of a moving elementary particle. The particle ionizes atoms along its path which causes rapid localized boiling and thus bubbles. The bubbles are large enough to photograph and thereby record the track.

quantum-mechanical in nature. So, according to von Neumann, only one step in the measurement process is different from all the others: *the mind of the observer*. Rooted as he was in the classical tradition, von Neumann came to this conclusion most reluctantly.

Von Neumann's approach was carried to its logical conclusion by Eugene Wigner, who introduced the brain of a sentient observer into the measurement system. The brain is in a superposition of states represented by some overall wave function. The mind associated with the brain makes its selection, and one of the many potential states is realized. Thus, to create the physical world, a wave function must be collapsed by a nonphysical entity, which Wigner terms sentient consciousness.

Bohm suggested that a mindlike quality exists in all quantum states, and that is what creates the nonlinearity. (We shall treat Bohm's ideas in more detail in Chapter 14.) On the other hand, Roger Penrose views the collapse as a gravitational phenomenon.[4] According to all approaches, the rules of linear superposition require an outside agent to create a nonlinear situation. Scientists and philosophers have expressed much concern regarding the characteristics of this agent since in general they are resistant to the notion of introducing a nonmaterial instrument such as mind.

Where does that leave us concerning the wave function? First, it describes a sea of possibilities, which we call a superposition of states. Second, an agent of some kind is required to make it nonlinear, that is, to select a particular state to be actualized.

---

[4]Penrose makes the assumption that whatever is occurring on the subatomic level must eventually be registered on the macro level. In this process, a single quantum of gravity (a graviton) is created, producing a new nonlinear situation which thereby breaks the superposition principle.

7 Phase entanglement, or interconnectedness

Quantum theory tells us that the universe does not consist of separate parts but appears to be one seamless whole. Perhaps a better way of saying it is that the universe is not made up of discreet, precisely defined elements. Extracting elements for examination can only be done in a limited way, because in some sense, everything is tied to everything else. The aspect of quantum theory that leads to this concept of wholeness is called phase entanglement.

Phase refers to the stage in the cycle of a periodic wave. In a wave on a lake, any point on the water moves up and down in relation to time, and at any given moment the wave is at a certain phase in this cycle. When two waves are "in phase," their crests and troughs coincide. When they are not in phase, there is a certain phase difference (or phase angle) between them. On a complex plane, when a wave function has a projection on a particular axis and thus creates a second vector, the angle between the two vectors is the phase difference at any point in time.

We have already discussed the linear superposition of different quantum waves for the same system. But suppose we have two or more systems. Two or more quantum systems can share the same quantum wave. When this happens, we say that the systems are entangled with each other. The separate systems no longer have quantum waves of their own. This process "quantum connects" the two systems no matter how far apart they get from each other, provided that they are protected against collisions with other particles. Collisions with other particles in the environment can break the quantum connection apart.

To clarify this, let's suppose that both systems are described by one overall wave function. Therefore, the first quantum system does not have a wave function of its own, and, neither, of course, does the second. The separate systems have lost their individual quantum wave identities and have merged into a discarnate whole larger than the sum of its parts. If the two systems separate, the overall wave function need not be destroyed, though it can be. Rather, we can say that once two systems are entangled, the overall wave function can provide an available instantaneous connection. This is what we mean by phase entanglement.[1] Here is Herbert's eloquent description of this process:

> Since there is nothing that is not ultimately a quantum system, if the quantum phase connection is "real," then it links *all systems that have once interacted at some time in the past* . . . into a single waveform whose remotest parts are joined in a manner unmediated, unmitigated, and immediate. The mechanism for this instant connectedness is not some invisible field that stretches from one part to the next, but the fact that a bit of each part's "being" is lodged in the other. Each [quantum particle] leaves some of its "phase" in the other's care, and this phase exchange connects them forever after. What phase entanglement really is we may never know, but Bell's theorem tells us that it is no limp mathematical fiction but a reality to be reckoned with. (Herbert, *Quantum Reality*, p. 223)

The word "instantaneous" sends up a red flag to the physicist. An instantaneous connection through the three-dimensional world violates the spirit of relativity if not the fact. But keep in mind that the overall wave function is not in three-dimensional space but in a multidimensional space. Consider this scenario: two electrons come together in some interaction and then part, going their separate ways. If we observe the first electron, it will have changed its

[1]This connection was recognized by Schrödinger in the early days of quantum theory. Since the entire universe can in principle be described by an overall wave function, all the particles in the universe are phase-entangled.

characteristics, which is to say we will have made some undefined potentiality a fact in our world. The amazing thing is that our observation instantaneously changes the characteristics of the second electron as well.[2] Now, in our three-dimensional world, we do not have any means of passing information faster than the speed of light. So, the change must be happening outside the three-dimensional world, that is, in the domain of the wave function. It might be suggested that this faster-than-light linkage exists only in the mathematical formalisms of quantum theory and is not necessarily real. To examine whether this is the case, we turn to a theorem developed by Bell. The theorem and its experimental verification indicate that this superluminal connection actually exists; that is, its effects are real in the three-dimensional world.

To understand Bell's theorem, we need to go back to Einstein's discussion with Bohr regarding the interpretation of quantum facts. Even though Einstein's work contributed to the success of quantum theory, he felt the theory was incomplete, largely because he believed strongly that the world must be viewed as both real *and* consistent with quantum facts. His debate with Bohr was carried on through "thought experiments" — logical problems which are created mentally but not carried out practically as experiments. The aim of Einstein's thought experiments was to convince Bohr that quantum theory was incomplete, that some essential element was missing. The most famous of these was an experiment Einstein coauthored with Boris Poldosky and Nathan Rosen; it is known by their initials, EPR.

---

[2]Bohm supposes an arrangement where pairs of electrons are shot out in opposite directions. The initial structure is such that if the spin of one electron points up,the other electron must have a spin pointing down. When they are separated by a distance that precludes a light connection, then according to relativity theory, if the spin of the first electron is changed, the second should not be affected. However, quantum theory tells us that the second electron spin will change accordingly. This is accomplished through phase entanglement.

In discussing the EPR experiment, we shall use the terms *local* and *nonlocal.* As stated earlier, these terms define the causal relationship between events. In a local reality, information cannot be transmitted faster than the speed of light; in a nonlocal reality, events can influence each other faster than light, or instantaneously. The classical principle of retarded causality is the "common sense" axiom that causes must always happen before their effects in all frames of reference. This logically independent principle, when combined with special relativity, requires that reality be local. However, all of the experimentally confirmed predictions of special relativity, like time dilation and the equivalence of mass to energy, remain theoretically intact, even if reality is nonlocal, thus violating the principle of retarded causality.

The EPR thought experiment proposed the following. Two particles are moving relative to each other. Thus, they have momentum, and there is a certain distance between them. While Heisenberg's uncertainty principle states that the position and momentum of a particle cannot be measured simultaneously and with certainty, the total momentum of the particles and the distance between them can be measured with certainty.[3] Now, if the two particles interact and then fly apart at near the speed of light, they may end up at opposite ends of the galaxy.

Suppose we station a physicist on one side of the galaxy to measure the momentum of the particle there. Since momentum is conserved, he automatically knows the momentum of the other particle. If he measures the position of his particle, the position of the other particle is also known. It is true that measuring the position of

[3]Suppose you know the total momentum of both particles. A light ray with a very short wavelength sent from one particle to another would give the distance between the two particles but it would disturb the momentum of each particle. However, the total momentum would not change. Therefore, the total momentum and the distance between the particles can be known simultaneously and with certainty.

the particle made the momentum uncertain, but this should not affect the second particle since it is too distant to be changed by a signal traveling at a rate less than the speed of light. If you accept this locality requirement, as Einstein did, the physicist doing the measuring determined with certainty the momentum and position of the second particle without disturbing it. This in effect violates one of the basic tenets of quantum theory, the uncertainty principle. So, according to Einstein, either relativity's locality restriction is incorrect or, in some way, quantum theory is incomplete. Since relativity theory had never been found wanting in any experiment, Einstein concluded that the latter was the case.

In responding to the challenge, surprisingly Bohr did not attack the locality requirement. Rather, he concluded that Einstein's error was in the assumption that particles have well-defined positions and momenta without a prior observation. In short, there is no conflict with relativity theory *if* one accepts that the two particles are still part of a unitary quantum system and have a single common wave function even though they are separated to the far corners of the galaxy. Bohr agreed that no physical signal or influence could travel from one particle to the other, but felt that one cannot ignore the experiments carried out on one particle when discussing the attributes of the second particle. In his opinion, even if no physical force is involved, both particles will cooperate to a certain degree in their behavior. Basically, according to Bohr, they can no longer be regarded as independent real particles because of phase entanglement. Einstein could not accept this idea, which he called "spooky action at a distance."

Bohr and Einstein disagreed profoundly on this question, but no resolution was forthcoming until 1965. In that year, Bell published a mathematical theorem that was later used to decide the question in the laboratory. Bell theoretically investigated the problem of a two-particle quantum system such as we have been describing. He began with the assumption that Einstein was correct. That is to say, all the forces involved are classical and cannot be transmitted faster than light. In essence, the dynamic attributes of the two particles are always well-defined — no latent potentialities are allowed. Through that assumption, he established a limit on the possible level of correlation for the simultaneous measurement results on the two particles. This limit became known as Bell's inequality, which could be tested in the laboratory if the experimenter were sufficiently clever. If the limit is violated, quantum theory is correct and connections faster than light can take place. If not, Einstein's views would be valid.

A number of experiments were performed, and quantum theory seemed to be validated. There was still room for doubt, however, until 1982, when Alain Aspect performed what is considered the definitive experiment.[4] The result — Bell's inequality was violated, and quantum theory was shown to be correct. That is, our real and local world is immersed in and interpenetrated by a more basic reality that is invisible and nonlocal. In short, whatever else the domain of the wave function may be, it is certainly a seamless whole since it is definitely nonlocal in character. That being the case, we cannot be the kind of biological machines described by the classical physicists — a gestalt of atoms, molecules, and cells subject to local forces only. With Bell's theorem, we have opened the door to the enigmatic world of nonlocal correlations.

---

[4]While Bell developed his theorem in the mid-1960s, a practical test of the inequality could not be undertaken immediately, since the technology was not yet available. To determine that two particles cannot communicate with each other by conventional means, the time intervals must be extremely short, to be sure the particles are not influencing each other with signals of light speed or less. Some early attempts were made to test the theory but their accuracy was open to question. However, Aspect's experimental arrangement using the polarization angle of two oppositely moving photons emitted from the same atom was refined enough to convince all physicists.

Earlier we suggested that both quantum theory and relativity theory point to the fact that the universe is an indivisible and unbroken whole. Quantum theory certainly endorses wholeness through the concept of phase entanglement. Relativity theory also has a nonlocal feature which is demonstrated in a different way. To understand this, let us return to the concept of inertia, the property of a body that tends to make it stay in a state of rest or uniform motion in a straight line unless acted upon by an external force. Now imagine that there is just one particle in the universe. What could motion mean in a situation where there would not be a reference frame against which motion could be measured? The particle obviously could not have inertia as defined by Newton. Now suppose there is a second particle. Would that create the inertia in the first particle? This question has not been unequivocally answered. Many physicists, however, endorse the concept that a particle's velocity is determined against the position of all the matter in the universe.[5] Therein lies the problem. If the particle instantly knows to change its motion relative to all the matter in the universe, then nonlocality is a feature of general relativity as well as quantum theory.

Bohm uses an analogy that helps us understand the transition from separate elements to undivided wholeness. He says that our notion of order is based on our perceptual experience. The development of the lens has led to such devices as telescopes and microscopes which allow us to perceive objects beyond the range of the naked eye. That experience, in turn, led us to the conception that reality consists of separate elements and encouraged the mechanistic way of thinking. This view not only pervaded science but entered every phase of our lives. Bohm asks, "Have any instruments been developed that would similarly help point vividly to the way of

---

[5]The idea that the inertia of a body is dependent on all the mass in the universe was proposed by Ernst Mach in the nineteenth century and is known as Mach's principle.

8 Time and the wave function

Newton saw time as an absolute and linear flowing entity. Even though this assumption placed his laws on a firm mathematical footing, a problem remained: the direction of time. When velocity and acceleration are described mathematically, the equations do not distinguish between time running forward and time running backward. That is, whether we use plus *t* or minus *t*, the equations are the same. This is known as time-symmetry, and Newton's equations describe a world that is time-symmetric, or time-reversible. In our daily lives, of course, we have no experience of such reversibility. We may have seen a glass fall off the table and shatter, but we have never seen the pieces put themselves together and return to the table intact. Yet Newton's laws say that this is possible. Interestingly, when we examine Schrödinger's equation with respect to time, the same reversibility is present. That is, physicists agree that quantum theory is time-symmetric before a potential state (described by the wave function) becomes actual.

This difficulty is supposedly resolved for Newton's equations by the second law of thermodynamics. According to this law, in any physical process, such as the smashing of the glass, energy is wasted as heat, and perhaps sound, and can never be recovered. This loss of energy is equivalent to an increase in the quantity called entropy. Entropy is therefore linked with time in a way that maintains the evolution of the universe in a forward direction — indicated by an arrow — and increasing entropy.[1] So Newton's mechanics are time-

[1]Entropy is the measure of the energy of a system that is unavailable to do work. In any irreversible change in a closed system, the entropy increases. Even though the total energy of a closed system has not changed, the available energy is less. Therefore, the higher the entropy, the more disordered the system. The whole concept of entropy is a consequence of the second law of thermodynamics which deals

less, but the second law of thermodynamics is not. How can the two be reconciled?

The solution came from Maxwell and Ludwig Boltzman in the form of the kinetic theory of gases. Basically, Boltzman envisioned a gas as a tremendous number of molecules in ceaseless motion, colliding with each other and the walls of the container in a chaotic way. The pressure inside the container corresponds to the impact of the molecules on the walls, and the temperature of the gas corresponds to the motion of the molecules. All this can be expressed mathematically. To picture the increase in entropy over time, imagine that a system starts out in a fairly orderly way, with the faster molecules to the left and the slower molecules to the right. Therefore, the temperature will be higher on the left than on the right. If the system is isolated and undisturbed, the faster molecules will collide with the slower ones, and eventually the temperature will be evenly distributed. This indicates a decrease in available energy, or an increase of entropy with time. The concept of entropy makes time asymmetric, flowing in the forward direction only.

At first glance, reversing the direction of this change seems impossible. But actually such a reversal is only highly unlikely. The state the gas is in at any time can be designated by specifying the position and momentum of each molecule. With time, the gas changes from one state to another. Most states are largely chaotic and cannot be distinguished because in practice we cannot account for each molecule. So, if a gas starts out in an orderly state, it is overwhelmingly probable that subsequent states will be less orderly because disorderly states far outnumber orderly ones. For example, a pot of water over a fire is *almost* sure to boil rather than freeze. But in fact, for the individual molecule, there is no arrow of time, and

---

with the direction in which any physical or chemical process involving energy can go, a direction that is correlated with the direction of time. If the universe is a closed system and entropy is always increasing, then time always goes forward.

Newtonian laws are followed. For a large conglomeration of molecules, however, only averages can be measured; these are subject to the law of increased entropy, which introduces an arrow of time. In practice we do indeed see that the heat is distributed more evenly, and entropy increases.

While in classical physics particles can be considered similar to little billiard balls, particles in quantum theory cannot be so described. That leaves us with the question: Does quantum theory have an arrow of time? The answer is yes: while relativity is a time-symmetric theory, quantum theory is not. Recall that the Schrödinger wave equation describes how an initial state develops into a number (any number) of subsequent potential states as time progresses. As we have seen, it is a deterministic equation, similar to classical equations of the same kind. On the other hand, the Copenhagen interpretation specifies that upon observation only one state is selected, which then enters our real world. The probability of a particular state actually occurring is given by a second equation. The two equations are connected only by an act of observation; the second equation is not deduced from the first. The probability equation is a separate procedure that has been added to quantum theory to obtain agreement with experimental fact.

The Schrödinger equation is time-symmetric, and the probability equation is not. Since there seems to be some confusion about this in the literature, I would like to quote Penrose on the subject. (R refers to the probability equation, U to the Schrödinger equation; squaring the modulus means multiplying the quantum complex amplitude by its complex conjugate.)

> . . . The rules for the R part of quantum mechanics simply cannot be used for . . . reversed-time questions. If we wish to calculate the probability of a *past* state on the basis of a known *future* state, we get quite the wrong answers if we try to adopt the standard R procedure of simply taking the

> quantum-mechanical amplitude and squaring its modulus. It is only for calculating the probabilities of *future* states on the basis of *past* states that this procedure works — and there it works superbly well! It seems to me to be clear that, on this basis, the procedure R cannot be *time-symmetric* (and, incidentally, therefore cannot be a deduction from the time-symmetric procedure U). (Penrose, p. 359)

What can we conclude from this interesting distinction? It seems fairly clear that if we are describing the domain of the wave function, then the arrow of time is missing. When we introduce a chosen alternative into this world, the concept of time comes with it.

There is one more point I would like to leave with you. The second probability equation works beautifully in investigating procedures going from the past to the future. From this we conclude that the future is open. Surprisingly, if we go back in time, we find that past is open in the same manner: we cannot go back into a fixed path. Thus it seems as if the present moment is the only point of reality, and *both* the past and the future are open and undetermined. We will discuss this fascinating idea further in relation to Seth in Chapter 15.

The significant thing to remember about the wave function and time is that before collapse, the wave function represents a timeless order, a domain that does not have an arrow of time. However, upon the collapse of the wave function, this timeless domain is changed into the timed order of our three-dimensional universe. In physics, this is called the quantum-mechanical arrow of time and is the result of breaking the symmetry of the timeless order of the wave function. By our observations, we literally create an arrow of time.

9 Energy and the wave function

As we have noted, the idea of an electron as a particle similar to a small billiard ball does not fit with relativity theory or with quantum theory.[1] It was replaced with the concept of the electron as a structureless point carrying a negative charge.

All charged bodies are associated with an electric field which decreases with distance from its center, in accordance with the inverse square law; that is, as the square of the distance from the charge increases, the strength of the electric field decreases. But what happens if we go in the other direction? If the electron is a sphere, the electric field strength is greatest on its surface. If we let the radius of the sphere decrease to a point, the electric field strength goes to infinity. Since Einstein's equation states that $E=mc^2$, an infinite energy means an infinite mass. To physicists this paradox was impossible to deal with. To further confuse the issue, the *observed* mass of an electron was shown to be finite. James Jeans once noted that it is as meaningless to discuss how much room an electron takes up as it is to know how much room a fear, an anxiety, or an uncertainty takes up (Cole, pp. 204-205). As we proceed to examine these perplexing problems, we will find Jeans's comparison very apt.

One way to deal with this puzzle is to view the electron as having two parts — one the "bare" electron, ignoring the charge, and

---

[1]A large body like the earth is held together by the force of gravity. But an electron is too small to be subject to gravitation, and since it is negatively charged, its parts are subject to electric repulsion rather than attraction. What, then, holds the electron together? If it were rigid, and we strike it, the whole electron would fly off in some direction. That is, the force on the side that is struck would have to be transmitted instantaneously to all other parts if the electron is to move without changing its shape. But this violates relativity theory since we know signals cannot be transmitted that fast. On the other hand, if the electron is not rigid, why doesn't it blow apart by its internal repulsive forces? Partly because of this contradiction, physicists abandoned the ball idea.

the other the electric field itself. In practice, the bare electron cannot be separated from its field, but thinking of it in this way helps visualize the solution physicists developed for the problem. Called "renormalization," it involves infinite energy that is merely subtracted from both sides of the equation. You were probably taught in school that this procedure is not allowed. But if one knows the answers ahead of time through experiment and if this procedure is undertaken in a consistent, rigorous way, the results are remarkably accurate. Feynman, one of the physicists who discovered this method of eliminating infinities, called renormalization "a dippy process!" Resorting to such hocus-pocus, Feynman said, "has prevented us from proving that the theory of quantum electrodynamics is mathematically self-consistent" (Feynman, *QED*, p. 128). From a practical point of view, though, renormalization works, in that it gives values that are finite to both parts of the electron.[2]

The infinite field also can be related to Heisenberg's uncertainty principle, which states that the only way either the position or momentum of a particle can be measured decreases our knowledge of the other. Position and momentum are two incompatible observables, as are time and energy, to which the same principle applies. Since time is not strictly an observable, the best way to interpret the uncertainty relation in this case is to understand that there is a limit

---

[2]In his book *The Cosmic Code* Heinz Pagels has a wonderful analogy to explain the renormalization process. I can do no better than quote it:

> Suppose a man weighs 150 pounds on a bathroom scale. Then he has a good dinner and adds a couple of embarrassing pounds. But he decides to cheat by adjusting the bathroom scale so that it continues to read only 150 pounds. This cheating — or rescaling — is the renormalization procedure. If someone could actually put on an infinite amount of weight and then reset the scale by an infinite amount to give a finite weight, then one gets an idea of the amount of cheating or renormalization required in the calculational procedures of quantum field theories. Remarkably, for some field theories this cheating can be carried out in a mathematically consistent way, and these are called "renormalizable" quantum field theories (Pagels, p. 259).

to our ability to specify the amount of energy transferred and the time in which the transfer took place.

Because of this uncertainty, an electron can emit what are called virtual photons as long as they are reabsorbed in the proper amount of time to satisfy the uncertainty principle. Virtual photons come into existence out of nothing (even if the electron does not have the energy to emit them) and then fade away in an exceedingly short time.[3] There are instances when their effects are noticed, but they cannot be directly observed. With this in mind, envision the electron as emitting a great number of virtual photons, then reabsorbing them, according to the uncertainty principle. The result is a virtual cloud of energy surrounding the electron and decreasing with distance. However, like the electric field, the cloud of photons gets more energetic near the electron, approaching infinite energy at the center. In short, since the journey of a given virtual photon can approach zero, its energy cannot have a limit. When you take into account all the photons, the cloud can have an infinite energy. This is a quantum-mechanical description of the electric field we discussed earlier.[4]

Where does all this leave us? When we examine the particle aspect of the wave-particle duality, we find that physicists are forced to assume a point particle for its description. Mathematically, this gives us a near-infinite mass, but experimentally the best estimate for the mass of an electron is 9.1×10^{-28} grams. That is definitely not infinite — so is the calculated infinite mass just an anomaly of the

---

[3]To understand what is meant by the term "nothing," one must recognize that the vacuum, or empty space, is not really empty. It is actually a dense sea of virtual particles made of energy that exists for us only an extremely short time. That is, energy is borrowed from a pool of energy for a short period but the pool is normally not measurable by any of our instruments. So by "nothing" we mean that which cannot be measured. This concept will be clarified as we proceed.

[4]Actually, as you get nearer the point electron, virtual electron-positron pairs appear in addition to the virtual photons. In fact, eventually there are virtual particles of every kind.

mathematics, or does it have significance? The definition of energy is the capacity to do work.[5] The only way energy can be detected is through instruments and our sense organs. In some way, the energy must be capable of registering its existence through work. With that as a criterion, the infinite energy revealed only in mathematics cannot be in our three-dimensional world. To understand the nature of this mathematical energy we must turn to the concept of the vacuum.

Relativity theory replaced the vacuum of space with the geometric curvature of space-time. Wheeler looked at space-time from a quantum-mechanical point of view and saw it as a chaotic foam with a tremendous energy density. This foam was not smooth but grainy in texture because of random quantum jumps. The uncertainty principle implies that a great amount of energy can be borrowed from "nothing" if it is paid back in time. Since gravity is extremely weak on the quantum level, a great amount of energy must be paid back in very short times, creating distortions and bumps in space-time and, if the time is small enough, a labyrinth of holes and tunnels. What can it mean to have a hole in space-time? To avoid violating classical laws, which pertain to our universe only, we might consider these holes to be located at the boundaries of space-time, connecting to regions outside it via tunnels. In short, our universe can be seen as infested with holes and edges.

Another way of conceiving our universe involves the idea of quantum black holes.[6] In 1913, Planck established the relationship

---

[5]The work accomplished is defined as the product of the force required to move a body and the distance through which it moves. For example, when you lift an object that has a certain weight, you supply a force that overcomes the force of gravity through the distance it is lifted. The work you have done on the object gives it the ability to do the same amount of work; thus, it is called potential energy. The motion of a thrown ball, which can be used to knock something down, is called kinetic energy.

[6]A black hole is a material body that has collapsed to a highly compressed state in which the gravitational field is so intense that not even light can escape from its surface. It is characterized by only three properties: mass, charge, and angular momentum. (Hawking has shown that quantum effects do permit some radiation).

between the three basic constants of nature: the speed of light (c), Planck's constant (h), and the gravitational constant (G).[7] By allowing these constants to assume the value of one, Planck was able to create absolute units. That is, he combined the constants of gravity, relativity, and quantum theories and created natural units of time, length, and mass. The time is 5.3×10^{-44} seconds, the length is 1.6×10^{-33} cm, and the mass is 2.2×10^{-5} grams. For example, one second of time would contain 1.9×10^{43} absolute units, and one centimeter of length would contain 5×10^{32} absolute units. If we imagine a particle with these dimensions, it turns out that it would be a black hole of quantum-mechanical size.[8] These holes have not been created in the laboratory since enormous energy would be required, but a number of physicists consider them to be the basic constituents of space-time.[9] These are virtual black holes which are

---

[7]Planck's constant gives the ratio of the energy carried by a photon to its frequency. It appears in a number of relationships where some observable is quantized, meaning that the observable can take only specific separate values rather than any value in a given range. G is a fundamental constant which gives the proportionality in Newton's universal law of gravitation between the masses involved and the distance between them.

[8]The way to see such a quantum black hole is through Heisenberg's uncertainty principle. At the space scale of Planck's length, this small region of space-time can borrow enough energy to curve space-time back on itself, thus creating a mini-black hole. The payback time is, of course, Planck's time.

[9]Michael Green says:

> One way of describing quantum mechanics is in terms of the so-called uncertainty principle and by using the uncertainty principle, it's easy to argue that the shorter the distance scale that you're trying to describe, the more uncertainty there is in the energy of what you're trying to describe. Now in a theory of gravity this means that when you try to describe things at incredibly short distances (and by short I mean *unbelievably* short compared to the size, say, of even the photon), the fluctuation in energy of what you are looking at might be big enough to make a little black hole. So if we are contemplating making an observation on small enough distance scales (this scale is called the Planck distance . . .), we are forced to think of even empty spaces as consisting of an infinite sea of fluctuating black holes, coming and going in very short times. This of course radically alters our notions of what we mean by space . . . (Davies and Brown, p. 123).

popping in and out of existence at a rate equal to the Planck time. If we accept all that, then space is constructed of these black hole spheres tightly packed together, each sphere with a radius equal to 1.6×10^{-33} cm. So in one cubic centimeter, we would have 10^{99} black holes, and if each hole has a mass of 10^{-5} grams, then space has a density of 10^{94} grams per cubic centimeter. Here is how Wheeler states it:

> The enormous factor from nuclear densities, 10^{14} g/cm^{3}, to the density of field fluctuation energy in the vacuum, 10^{94} g/cm^{3}, argues that elementary particles represent a percentage-wise almost completely negligible change in the locally violent conditions that characterize the vacuum. [A particle (10^{14} g/cm^{3}) means as little to the physics of the vacuum (10^{94} g/cm^{3}) as a cloud (10^{-6} g/cm^{3}) means to the physics of the sky (10^{-3} g/cm^{3}).] In other words elementary particles do not form a really basic starting point for the description of nature. Instead, they represent a first order correction to vacuum physics. That vacuum, that zero order state of affairs, with its enormous densities of virtual photons and virtual positive-negative pairs and virtual wormholes, has to be described properly before one has a fundamental starting point . . . . (Weaver, Vol III, p. 681)

It should be emphasized that physics at present does not have a quantum theory of gravity. Therefore, all of the above conjectures are extrapolations from current theory and are not as yet proven. Nevertheless, it would seem that quantum theory strongly indicates that the vacuum contains an enormous amount of energy. This energy is ignored by physicists because it cannot be measured. If we choose not to ignore it, then matter can be seen as a tiny fluctuation of this huge energy domain, like a small wave on a vast ocean. But while matter is in space-time, this ocean of energy is not.

Bohm identifies this ocean with what he calls the implicate order and the waves on it with our three-dimensional universe, which he calls the explicate order. Briefly, Bohm sees the implicate order as enfolded or unmanifest while the explicate order (our world) is unfolded or manifest. So we can view a particle as manifest energy, folding in and out of this enormous ocean, rather than as a localized entity moving from one place to the other. Again, just like a wave on water, while the disturbance is propagated laterally, the water itself moves up and down, with the wave folding in and out of the water. Since our world consists of waves on this endless ocean of energy, we do not notice the ocean itself any more than a fish notices the water in which it swims.

As we noted previously, all gestalts of matter have an associated wave function — even the universe as a whole. All these wave functions can be seen as forms within the implicate order. Originally we stated that the wave function does not contain energy, but that was because as a complex function it cannot be measured. In fact, the wave function describes an energy pool that is vast, perhaps infinite, but unmeasurable — an ocean of probabilities and potential that may be the source of energy for the world of our experience.

10 Motion and the wave function

Quantum theory does not provide us with a clear conception of motion. According to the Copenhagen interpretation, the physicist can deal only with a series of observations, which involves a collapsing wave function at each observation. There can be no knowledge of what happens between observations. Wheeler describes "what we call reality" as "a few iron posts of observation . . . [filled in with] an elaborate papier-maché construction of imagination and theory" (Wheeler and Zurek, p. 194). In some sense, either the physicist or a measuring instrument must be present to produce what we call reality.

As an example, let us consider an electron in an atom. If we are not looking, a wave propagates around a nucleus with a mean radius determined by the energy level of the electron in question. This condition continues as long as the electron does not gain or lose energy. Now, suppose the electron gains energy from an electromagnetic field. One might assume that the wave will go from the initial level to the higher level of energy by traversing the intervening space in a continuous and deterministic manner. Because the wave is a probability wave, not a real wave, its flow with time tells us that the probability is increasing that the electron will be on the higher level. But the energy levels are discrete, and the atom cannot hold a part of a quantum of energy. So the energy transfer is discontinuous and indivisible, even though the wave aspect is continuous and deterministic.

As you recall, the Schrödinger wave equation gives a continuous, causal prediction of what happens to the wave function in a probabilistic way, and the electron will be observed only on one level or another, never in between. This "quantum jumping" of the electron from one level to the other without passing in between was a source of great bewilderment for the early quantum theorists. In a conversation with Bohr, Schrödinger said, with considerable frustration, "If we are still going to have to put up with these damn quantum jumps, I am sorry that I ever had anything to do with quantum theory" (Moore, p. 228). To Schrödinger, the quantum jumping destroyed the space-time conception of atomic processes. That is, particles jump from one point to the other without passing in between; they do not pass through space-time.

The concept of motion is difficult even without considering quantum processes. To represent the movement of an automobile, we may employ a series of different positions, as in the frames of a movie. However, the frames alone do not convey the sense of the auto continuously covering space as time progresses. We get around the problem in a movie by quickly flashing the sequence of pictures. We cannot reduce the time to zero because the car would then be stationary, but in a single exposure we can blur the position somewhat to imply motion, as in a photograph of a moving object taken at a slow shutter speed. In a movie, we assume continuous motion because of "visual persistence" in the rapid succession of static views; that is, the image is retained through the short periods between frames when the screen is dark. Similarly, Zeno noted that at any instant, an arrow shot through the air occupies a given place and is at rest at that point. Since the point of rest is arbitrary, the arrow cannot be moving at any instant, and therefore flight is illusory. Today we may conclude, based on our knowledge of quantum theory, that Zeno was not too far off. That is, an arrow going from point A to point B would go half the distance to B, then half of the remaining distance, and then half of that distance, so that its path is divided into infinitely smaller and smaller distances. No matter how small the remaining distance, one is left with the problem of how the arrow can get from A to B without some discontinuous jump. That is similar to the quantum jump; in quantum theory movement is discontinuous.

Another peculiarity of quantum theory is that electrons are absolutely identical. If an electron in my body is exchanged with one in your body, the state of my body and of your body would remain exactly the same. The same is true of atoms. We could even go so far as to say that if we exchanged every particle in your body for the same particles in my body, we would still remain exactly as we are. So, what distinguishes one person from another or one object from another is not the content but the form or pattern in which the particles are arranged.

All this indicates that the classical approach to motion is not adequate to explain the quantum facts. Instead of localized objects moving from one place to another through space-time, electrons when they are unobserved are not even discrete objects, although they have the potential to appear that way when a measurement is made. In between observations, electrons seem to be a flowing movement of probabilities — a movement that cannot be detected by instruments. Or, said another way, we have no direct sensory access to the domain of the wave function. This implies that there are at *least* two levels of reality. The first is our everyday three-dimensional world, and the second is a field model that is explained and comprehended by physicists only mathematically. As we mentioned earlier, Bohm attacks the problem directly by defining the two levels: the wave function domain (implicate order) is the unmanifest or enfolded part of reality, and our three-dimensional world (explicate order) is the manifest or unfolded portion of reality. When reality is unfolded, we experience what we call perception.

To return to our former metaphor, suppose we envision space as a vast ocean of energy and matter as a small wave on its surface.

Matter is what is manifest, and it has a relative stability, something like a cloud in the sky. The vast energy ocean, which Bohm calls the implicate order, is not in space-time, but space-time is a projection from it. That is, the ocean of energy is unmanifest, while the particle is manifest and represents a very small amount of matter or energy. (In physics terms, the particle is the quantized state of the field, but the field is spread all over.) Remember that as the original disturbance (e.g., a rock thrown into the water) produces a wave that travels across the water, the movement of the water molecules up and down is like a wave projected from the ocean and then injected back into it at each moment. This gives rise to the next wave, which in turn is projected and then injected back into the ocean. The energy of the wave moves outward across the ocean, but the water itself is projected and injected moment by moment.

Normally, a wave of water is composed of a mixture of waves of different frequencies and wavelengths. When a pulse is created, the shape of the pulse rapidly disperses because large waves travel faster than short waves in water. There are, however, pulses that are able to retain their form over rather long distances without dispersing. These are called solitons. Through a nonlinear interaction, the various wavelengths are coupled together and hold their form. So in one sense, the soliton is like a material entity, but in fact it is the movement of energy within a certain form. If a soliton moves from point A to point B, then the form is identical at the two points, but the content is not. In Bohm's terms, though the soliton is created and recreated moment by moment, its form is retained. The constantly changing content is probably why when two electrons interact there is no way of identifying which is which after they are separated. Electrons are less like billiard balls and more like solitons, which mix or merge and then part, retaining only their forms. Matter can be seen as just such a persistence of pattern.

The soliton is also useful in understanding several other concepts. First, since the soliton requires a nonlinear interaction to form, and the domain of the wave function is linear, something must introduce nonlinearity. To Bohm, it is the superimplicate order. To Wigner, it is sentient consciousness. To Bohr, it is a measuring instrument. To Penrose, it is a gravitational effect. All of which returns us to the measurement problem in quantum theory, which we shall discuss further. A second issue is how the soliton at A causes the soliton at B. Obviously the relation of the two cannot be through our three-dimensional world since the soliton disappears and reappears moment by moment, retaining only its form. The form and the connection between A and B are mediated by the body of nonmeasurable energy without a direct three-dimensional link. A third point is that if the ocean is considered as one whole and not subject to relativity, the so-called causation between A and B could very well have a nonlocal component. That is, space and time are not relevant in the implicate order. So in some sense, the implicate order is the genesis of form and is the vehicle for the nonlocal connections in quantum theory.

Motion in terms of quantum theory is a series of projections and injections from another level more basic than our three-dimensional world. Continuous motion as we perceive it is deceptive, just as a string of lights flashing on and off in rapid succession creates the illusion of movement. In neither case is there a localized entity traveling through space-time. In summary, we can say that while Newton envisioned bodies crossing space in a given amount of time, quantum theory tells us that motion is an enfolding and unfolding from the hidden domain of the wave function.

11 Tachyons and the wave function

We discussed Einstein's special theory of relativity in some depth in Chapter 2. In this section we shall focus on one particular aspect of that theory. In general terms, relativity theory indicates that as the velocity of an object increases, the size and mass change. Specifically, as the object approaches the speed of light, the mass increases and approaches infinity while the length approaches zero. These changes have been verified experimentally. In an atom smasher, a particle's mass is increased when it is moved at a velocity approaching that of light, which increases its force when it strikes a target. (This is the principle behind particle accelerators used to shatter the nuclei of atoms.)

When we talk about mass, we should be clear about the difference between inertial mass and what we ordinarily call weight. Weight is dependent on two elements, inertial mass and gravitational field intensity. When an object moves from one gravitational field to another, its inertial mass remains constant, but its gravitational mass changes with the intensity of the gravitational field. For example, on Mars your weight would be considerably less than on Jupiter because of the gravitational differences.

Inertia is an inherent property of matter. It is implied by Newton's first law of motion, which states that an object continues in a state of rest or uniform motion unless acted upon by an external force. That is, an object resists a change in motion by reason of its inertial mass. When an object is at rest, the inertial mass and gravitational mass are equal in a uniform gravitational field, as on Earth. When an object is placed in motion, its inertial mass increases. Since this mass is unbounded as it approaches light speed, quite often we hear that it is impossible to accelerate a mass up to or beyond the speed of light. While that is true, there is a loophole connected with the term "acceleration."

Basically, superluminal speeds cannot be attained by constantly increasing the force on an object. However, as pointed out earlier, Einstein's equations do not prohibit a particle from moving beyond light speed if that particle is traveling at a superluminal speed at its moment of birth. Since energy and mass are related by $E=mc^2$, a quantum particle can be created out of energy under the proper circumstances and could travel at superluminal speeds. Another way to attain such speeds involves the concept of tunneling. If a particle is traveling close to the speed of light and is allowed to tunnel through a barrier, then its speed could become greater than light.[1] We have assumed that when a particle is jumping, or tunneling, between two events, it literally is out of this world. Thus, we might say that Einstein's restrictions apply to our familiar three-dimensional world — that special relativity theory does not outlaw superluminal travel, but confines it to the right conditions. Relativity theory also says that superluminal travel can lead to time travel, in that superluminal signals sent between two events in our space-time can appear to some observers as going backwards in time. This idea is expressed in the following limerick:

> There was a young girl named Miss Bright,
> Who could travel much faster than light.
> She departed one day,
> In an Einsteinian way,
> And came back on the previous night. (Gamow, p. 104)

As a material object Miss Bright's body cannot go faster than light, but this is not true for the hypothetical particle we shall now describe.

---

[1]If a particle travels through a potential barrier and yet in classical terms does not have enough energy to pass through, this is called the tunnel effect. It can only be explained by quantum theory. In the process, the particle could achieve superluminal speeds.

Tachyons is the name given by Gerald Feinberg to hypothetical faster-than-light particles. There have been no observed phenomena that can be attributed to tachyons, but this does not mean that Feinberg's reasoning is incorrect. Similarly, the evidence for the existence of quarks is purely theoretical; no one has produced a quark in the laboratory. Yet the quark model is very useful, and the same may prove true for tachyons.[2]

Tachyons have some very interesting attributes. For one thing, they have a rest mass that is imaginary — our old friend *i* pops up again. The rest mass is the mass as measured when both the particle and the observer are at rest in the same frame of reference. Relativistic mass is the mass of a moving particle measured by an observer who is at rest and who is also in the same frame of reference.[3] Photons, which travel at the speed of light, have a zero rest mass; particles that do not have zero rest mass cannot reach light speed. Since tachyons have an imaginary rest mass and cannot be directly measured, most physicists ignore them. Nevertheless, this complex space keeps reappearing in the equations of physics. Could it be that the complex space actually underlies our space-time and is the primary source from which we generate our own space? There is some evidence to support this view.

A peculiar situation arises in considering the wave function for a tachyon, because the imaginary rest mass is used rather than the real rest mass of ordinary particles. When the wave function velocity for ordinary particles is calculated, it turns out to be faster than the

---

[2]Some string theories envision the vacuum as unstable and capable of exploding into an infinite number of tachyons. If a vacuum is considered to be zero energy, this just does not make sense. But, as we have seen, the vacuum contains enormous, unmeasurable energy.

[3]Discussing the rest mass of a tachyon seems somewhat contradictory because tachyons are never at rest. Actually, the significance of the so-called rest mass is that this property measures out to be the same number for all inertial observers moving uniformly relative to each other. Physicists often call the rest mass by other names, such as proper mass or invariant mass.

speed of light. Since the wave function represents the probability of observing a particular attribute, that is all well and good. But when we follow the same procedure for the tachyon, we get a surprising topsy-turvy situation: the wave function travels slower than light. Instead of having its wave aspects in an underlying world, it is as if the tachyon is from another domain, and its wave aspect is in the three-dimensional world.

Still another interesting property of tachyons has to do with the relationship between energy and velocity. Recall that so-called real particles speed up as they gain energy, which is why their relativistic mass increases with velocity. A tachyon does exactly the opposite: as it loses energy, its velocity increases. In fact, if it has zero energy, it will travel at an infinite velocity. But when energy is added to the tachyon, it slows down to our light speed. To reach light speed, an infinite quantity of energy is required. So the speed of light seems to be a dividing line between our everyday particles and tachyons.

To sum up, tachyons can be seen as the mirror image of normal particles, with the velocity of light as the mirror. As a tachyon gains energy, its velocity decreases. As it loses energy, its velocity increases. When all energy is removed, it travels at an infinite velocity and is at its rest mass. Some physicists call this a transcendent state. As we shall see, the concept of the tachyon is important when examining Seth's theories.

12 Consciousness and the wave function

Even though relativity theory was a radical departure from Newtonian mechanics, it is still basically a classical construct in that it describes a reality that is continuous (with no quantum jumps), deterministic, and local. At heart, Einstein remained a classical physicist. He saw the world as objective and physical and considered the immutable laws that govern its processes to be independent of human beings. In his view we are the audience watching a play unfold onstage. Although Einstein had a role in the development of quantum theory, he never really came to terms with it. He did not reject it, of course, since it was corroborated by experiment. But, as we mentioned before, in Einstein's mind, since quantum theory is capable only of making statistical predictions, it must be incomplete, and he felt that someday it must be encompassed by a broader theory. Also, the philosophical implications of quantum theory disturbed him greatly. The following excerpt from a letter Einstein wrote to Born demonstrates his misgivings.

> Bohr's opinion about radiation interests me very much. But I should not want to be forced into abandoning strict causality without defending it more strongly than I have so far. I find the idea quite intolerable that an electron exposed to radiation should choose *of its own free will*, not only its moment to jump off, but also its direction. In that case, I would rather be a cobbler, or even an employee in a gaming-house, than a physicist. (Holton and Elkana, p. 69)

As you remember, the quantum state is made up of alternatives, or possibilities that might occur. All these possibilities coexist, and, added together, they evolve with time. Each individual alternative has what is called a possibility amplitude, but these possibilities are not elevated to our physical world unless a measurement or observation takes place. When that happens, the quantum state "jumps" to one or another possibility. Quantum theory then requires a second equation, one which allows the physicist to compute the probability of the quantum state jumping to a possible alternative state. Note again that the connection between the two equations is a measurement or an act of observation.

If our entire reality is quantum-mechanical, what collapses the wave function? What changes probability to actuality? Since it must be nonmaterial, von Neumann decided it had to be the consciousness of the observer. The conscious mind thus becomes an essential participant in this process of bringing possibilities into actuality. In his book *At Home In The Universe*, Wheeler asks, "What is missing from the quantum story?" and answers as follows:

> Is not the central role of the observer in quantum mechanics the most important clue we have to answering the question? Except it be that observership brings the universe into being what other way is there to understand the clue? (Wheeler, p. 45)

A basic principle of classical physics is reductionism, which explains complex phenomena in terms of simpler ones. Reductionism denies the possibility of a collective property of a system that supercedes and cannot be explained by its component parts. In classical physics, then, the arrow of causation is from the individual to the aggregate. But with quantum theory the elementary particles are just potentialities until we introduce the laws of the aggregate. That is, quantum theory starts with the act of observation or measurement, which is a macro condition, and from this primary phenomenon, the elementary particle comes into being. The particle is literally not there until it is created by an act of observation. This downward causation is the opposite of reductionism in that the higher-level laws constrain and determine the activities of the lower levels (particles) — just the way it works in a living organism.

While Einstein was distressed by the notion of free will entering the realm of physics, other physicists have found the idea challenging and intriguing. In the words of Freeman Dyson:

> Quantum mechanics makes matter even in the smallest pieces into an active agent, and I think that is something very fundamental. Every particle in the universe is an active agent making choices between random processes. (Dyson, p.8)

If matter is in some sense conscious, it seems reasonable to bring the concept of information into the picture. And if we define causation as a movement of information, then the wave function can be said to represent information. When the information in the wave function is changed by an observation, then the electron changes its behavior — an example of downward causation modifying the physical world. So, assuming an electron makes choices, it does so with some constraints imposed by higher-level conditions.

Charon developed an interpretation of contemporary physics he calls complex relativity. His concept enlarges the field of physics from the study of matter to an equation in which Universe = Reality + Imaginary (sometimes referred to as Universe = Extension + Thought). Charon calls the imaginary side of the equation nonobservable phenomena, outside of space-time. This approach led him to the following conclusion:

> Complex relativity shows, and it is is one of its important results, that there does not exist in our universe what we previously called particles of "inert" matter. Any particle, whether stable or unstable, does possess what we may call a "body" (although eventually point-like) in reality; but this body is always associated with a nonlocal, nonobservable "head," made of one (and sometimes many) imaginary eons. To simplify, we might say of matter what

> biologists have for a long time been telling us concerning living organisms, from ants to human beings: they are made of a body associated to a conscious head. This unique privilege of living organisms is found at the simplest level of organization, the level of the individual particles of matter. (Charon, p. 53)

When Charon was asked whether a particle possesses an element of consciousness, he replied:

> Yes, that's it exactly. Complex relativity leads to a representation of a particle of matter that includes not only what it was considered to be before, that is, a part of matter, strictly speaking, but also a part that represents the mental structure of the particle. It is a little as if we described at the same time in this particle an "outside," which is the old representation of matter as we have known it, and an "inside," which had never been described until now, and which complex relativity allows us to represent. (Charon, p. 57)

Quantum theory brought the realization that physics can no longer be content with the study of only observable and "concrete" objects but must extend itself to consider objects outside our space-time that are nonobservable but do exist. Charon calls these imaginary objects — those which are not observable but are certainly representable — and concludes that physicists must now recognize that thought cannot be outside of the field of investigation of physics. To refer to the unity of the observable and nonobservable, the material and the mental, Charon coined the term "psychomatter."

Bohm summarized the concept of information in an interview with philosopher Renée Weber in which he describes the wave function as a kind of mental aspect of the electron. He suggested that when the electron reacts to the quantum field, it is somewhat analogous to a sentient being reacting to a situation. Now, if that sounds like Bohm is assigning at least a *minimal* consciousness to an electron, that seems to be just what he is saying. When asked if he were implying that an electron is alive, he replied:

> Maybe it is! What sense would it make to say it's not? The electron must behave in all sorts of strange ways, like being a wave and a particle at the same time and jumping from one state to another without passing in between — and doing all sorts of things that cannot be understood but only be calculated. If you don't want to say it's alive I suggest that you should say that the electron is a total mystery and all you can do is to compute statistically how it will reveal itself phenomenally in certain kinds of measurements. (Weber, p. 114)

All of this led Alastair Rae to say:

> Ever since the beginnings of modern science four or five hundred years ago, scientific thought seems to have moved man and consciousness further from the center of things. More and more of the universe has become explicable in mechanical, objective terms, and even human beings are becoming understood scientifically by biologists and behavioral scientists. Now we find that physics, previously considered the most objective of all sciences, is reinventing the need for the human soul and putting it right at the center of our understanding of the universe. (Rae, p. 67-68)

While we can perhaps accept the fact that an electron has some level of consciousness, how does that knowledge apply to the complex behavior of a human being? In short, how can we account for our obviously purposeful behavior? If we apply only the concepts of classical physics, each electron and atom in our body moves according to the forces (which are local) applied by neighboring atoms and molecules, and our purpose plays no role in the activity. For example, according to this view, when you walk across the room, your intent to do so is completely superfluous. Such an extreme mechanistic notion seems ludicrous; we know we could hardly take a step without some conscious intent. In short, mental events are certainly important. The question is, how are our thoughts — our consciousness — translated into action? Perhaps it is through phase entanglement and nonlocality that our intent becomes immediately available to all gestalts of energy, such as the electron, molecule, heart, and brain. Then, by way of their individual freedom and cooperation, the thought becomes action.

To help us visualize how this happens, let me repeat a metaphor I find useful. Imagine a symphony orchestra with a large number of musicians. Before the performance, the musicians are tuning or practicing with their various instruments. The sounds the audience hears are random, and it would be difficult to predict a musical composition from such a cacophony. However, when the conductor lifts the baton, each musician accepts a constrained role in the joint creation of the music. While a certain amount of leeway in interpretation is allowed for each performer, it is the symphony, or whatever composition is being played, that governs the whole. In other words, while the freedom of each musician is limited, a certain amount of individual creativity is allowed, which adds color and quality to the piece. In short, the musicians willingly cooperate for the common purpose of creating a performance, while still playing their individual roles in the process. The musicians are analogous to particles, and through downward causation — guidance from the conductor — certain purposes are achieved, resulting in fulfillment for both the parts and the whole.

Similarly, when you walk across the room, all your organs, molecules, and particles willingly cooperate in the action. Each part receives information from a higher level. Remember that all matter is described by a wave function, and all these wave functions are tied together through an overall wave function. Communication takes place through phase entanglement, which exists beyond space-time. For downward causation to work, all matter must be thought of as conscious at some basic level.

The following is a quotation from Herbert's excellent book *Elemental Mind*. In it he discusses various quantum interpretations, among them von Neumann's position. Herbert says:

> As a professional mathematician, von Neumann was accustomed to following his logical arguments boldly wherever they might lead. Here, however, was a severe test for his professionalism, for his logic leads to a particularly bizarre conclusion: that by itself the physical world is not fully real, but takes shape only as a result of the acts of numerous centers of consciousness. Ironically this conclusion comes not from some otherworldly mystic examining the depths of his mind in private meditation, but from one of the world's most practical mathematicians deducing the logical consequences of a highly successful and purely materialistic model of the world — the theoretical basis for the billion-dollar computer industry. (Herbert, *Elemental Mind*, pp. 156-157)

The idea of an agent, with whatever degree of consciousness, collapsing the wave function to select probabilities to become actualities, is the source of considerable consternation. Some physicists say the selections are entirely random — that as Einstein feared, God does indeed play dice with the world. If that is true, science cannot contribute anything meaningful regarding this domain. Other physicists believe the selection may be accomplished by physical processes as yet unknown, which they call hidden variables. Some scientists have placed consciousness in level three.[1] If that is the case, we need

---

[1]We shall explore the concept of levels in more detail later in the book. However, for the moment, consider level one as the domain of actual physical events. Level two is the hidden domain or the home of the wave function. All probabilities exist on this level. Level three is where the various probabilities are selected for actualization.

some theory or explanation for the interaction of consciousness and matter. Still another possibility is entertained by a few physicists — that not only is level three the home of consciousness, but levels two and one also contain consciousness. This point of view suggests that all energy — and therefore all matter — is conscious.

How can consciousness be introduced into physics? The standard approach is to envision consciousness as a phenomenon in its own right. The observation that rubbing certain substances together produced some mysterious and unknown forces led to the discovery of charge, and the observation that charge was transferred from one point to another led to the identification of electric currents, which in turn lead to the whole field of electromagnetism. Something similar might follow from observations involving consciousness, but the result would be a dualistic view of the world. That is, if consciousness is a separate phenomenon, then matter and mind are separate, and a theory is required for their interaction. A more promising approach is the transcendental point of view of Bohm: that is, matter and mind are aspects of one overall reality. Since consciousness, or mind, is not directly accessible to our physical instruments, then matter is the form that consciousness takes when introduced or projected into our everyday world.

13 Implications of the wave function

Interpenetrating levels and wholeness

We have noted that the wave function is linear, or passive, in that it requires an outside agent to select events from Heisenberg's potentia (level two) to take form as objective events in our world (level one). For us to enjoy the show we call physical reality, someone or something must act as director, and the need for such an agent implies yet another, or third, level of reality. That is, the first level is our three-dimensional world; the second is the domain of the wave function, housing all possible events that may occur in our world; and the third is that which selects among these events.

All levels are so intermeshed that we cannot consider one without including the others. This interconnection and interpenetration of levels is best expressed by the concept of wholeness. Basically, quantum theory states that the electron, or any object composed of elementary particles, does not have intrinsic properties, only incompletely defined potentialities. These potentialities are developed into "real" objects with intrinsic properties when the object interacts with an appropriate system. This refutes the classical view that the universe is made up of separate elements that interact according to exact causal laws, with the whole defined in a reductionist manner as the sum of its parts. In quantum theory, objects have properties defined only in interaction with other objects, and different interactions create different intrinsic properties. We now must come to terms with the fact that the universe is basically whole and indivisible, and we are enmeshed in innumerable underlying connections.

Physicists describe this wholeness by the concept of phase entanglement, in which quantum systems that interact are instantaneously interconnected. This nonlocal interconnection violates at least the spirit of relativity, and provides another good reason for assuming the existence of another level beyond level one: a second level is needed to establish the connection while still leaving the theory of relativity applicable in level one. Before Bell's theorem and Aspect's experiments, this interconnection was seen by some physicists as just a mathematical anomaly without any relationship to the real world. But we now know that Bell's theorem, which requires superluminal connections, is based on quantum fact, not just theory, and to dismiss phase entanglement as not real means disregarding experimental evidence. We can say with some confidence that the wholeness exhibited by the wave function redefines the mechanistic view of the universe as applicable only to a special case — level one — and therefore with limited validity.

The complex plane, bridging real and imaginary

In our excursion through complex numbers, we discovered that the complex variable is not only essential to the wave function formulation but shows up in relativity theory as well. One of the areas we examined was the time axis, or fourth dimension, in special relativity. Recall that Minkowski placed time on an equal footing with the three dimensions of space. However, because time requires a minus sign, when the square root is extracted, time becomes an imaginary number, written $-ict$. Of course, this imaginary aspect of space-time never shows up in measurements. But if an object moves at the speed of light, the complex number produces peculiar effects, or to put it more accurately, as far as the photon is concerned, space and time do not exist. So, in a sense, light seems to underlie level one and is described by complex numbers.

Complex numbers are also present in level two, which is described by the wave function. When Schrödinger first introduced his equation, a euphoria engulfed the physics community. Trained in handling water waves or sound waves in air, classical physicists saw the equation as a mathematical description they could understand. But it was not that simple. The wave function, which was a solution of Schrödinger's equation, was complex rather than real and so could not be accessed directly. But there was a way of obtaining real physical, and therefore observable, quantities. Multiplying a complex number by itself gives another complex number, so squaring the amplitude of a complex wave gives another complex wave rather than a real probability. But if the complex amplitude is multiplied by its complex conjugate, a real number results. This allowed the physicist to deal with level two, since by using this procedure, real predictions for the real world could be obtained from a complex world; or to put it another way, a complex space could be projected into a real space. This procedure seems to be telling us something significant about reality. Cramer used the same approach to develop his transactional interpretation (see pp. 73–74), in which the quantum interaction requires a two-way handshake utilizing both retarded and advanced waves. We saw that the advanced wave goes back in time, so in some sense the future affects the past. Wolf relates this concept to the wave function and its complex conjugate, an idea we shall run across again.

Before we leave this review of complex numbers and the wave function, let us see how Pauli interpreted the complex plane in psychological terms. Pauli was a good friend of Carl Jung, one of whose great achievements was the description of universal archetypal symbols present in the collective unconscious.[1] Pauli entered into analysis with Marie-Louise von Franz, one of the more famous

[1]Archetypes in the Jungian sense are universal primordial images passed down to us from our ancestral past, including prehuman and animal ancestors. They are not part of conscious thought but serve as predispositions to certain behaviors. They can be linked to such instincts as fear of the dark and the maternal instinct.

disciples of Jung. Under her tutelage, Pauli employed a process called active imagination, a technique of entering into direct dialogue with aspects of one's unconscious. In one of these encounters, he discussed wholeness with a Chinese woman who personified Pauli's anima (the feminine aspect present in men). At the close of the conversation, the Chinese woman slipped a ring from her finger. As the ring floated in the air, Pauli recognized it as the complex unit circle, a form created on the complex plane. Figure 4 shows this: a vector one unit long is rotated counterclockwise from the real axis, tracing a circle with a unit radius; after one-quarter turn, the radius is *i*; after one-half turn, the radius is minus one, and so on. In Pauli's active imagination, he and the Chinese woman call this circle *i*.

> Pauli: The *i* makes the void and the unit into a couple.
>
> Woman: It makes the instinctive or impulsive, the intellectual or rational, the spiritual or supernatural, of which you spoke, into the unified or monadic whole that the numbers without the *i* cannot represent.
>
> Pauli: The ring with the *i* is the unity beyond particle and wave, and at the same time the operation that generates either of these. (Laurikainen and Montonen, p. 353)

The dialogue quoted above was published in a paper by Dutch physicist Herbert van Erkelens, who wrote about Pauli and his life. Van Erkelens comments that "the imaginary unit is the expression of a mathematical structure in the unconscious. It is a number archetype." In the active imagination dialogue, Pauli is indicating that in some sense the complex plane unifies the particle and the wave as aspects of one whole. It is only when we display the particle aspect for our own purposes that a split occurs.

To use quantum theory to describe experiments in the three-dimensional world, we must leave the domain of complex numbers

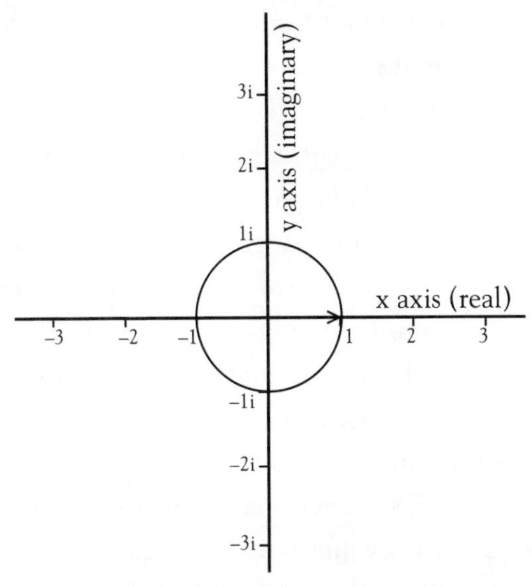

Figure 4. Complex unit circle

and return to the "real" world; then we can make choices among contradictory (or, in Bohr's terms, complementary) states. The act of observation projects the quantum state vector in Hilbert space to one of a set of mutually perpendicular axes that forms a basis or a frame of reference. So in a beautiful way, complex numbers allow us to represent both observable and nonobservable phenomena, describing the imaginary realm of level two and also providing us with a mathematical formalism that allows us to make real number predictions. One cannot avoid the notion that the concept of complex numbers is fundamental to understanding our world. In representing both the material and mental sides of matter, the complex number offers a way of comprehending how matter and consciousness are aspects of the same reality.

Energy and the manifestation of matter

In discussing energy in relation to the wave function, we found that a distinction had to be made between the manifest and the unmanifest. The manifest world that we directly interact with is measurable with our instruments, and the hidden second level is unmeasurable.[2] Often physicists define as real that which is measurable. Of course, that is not entirely correct since there are particles considered real that cannot be seen (measured) in bubble chambers. Perhaps their unmeasurability is ignored because these particles fit neatly into the reductionist scheme. But, in any case, generally, the rule is that if it contains measurable energy, it is real.

To understand this concept, let us return to our imaginary universe of two dimensions populated by two-dimensional people, "flatlanders." If they are technically advanced, they will be able to measure all components of energy involving the two dimensions of a flat plane. But any energy directed along a third dimension perpendicular to the plane will be unmeasurable. Perhaps its effects could be ascertained, but there would not be direct access for measurement. Analogously, our instruments measure only our three dimensions. However, if we consider the mind as an instrument, we might learn to measure other dimensions through altered states of consciousness.

The wave function, since it is complex, does not contain manifest energy. But as we have seen, an electron has a surrounding field of energy which mathematically computes to be infinite. Through renormalization, the infinities are removed so that the computational results agree with experiment. In a mathematical sense, the field is separated from the bare electron, though practically this cannot be done. If the electron is a point, it is surrounded by an infinite energy field, and if it is larger than a point, relativity rules

[2]Actually, this unmanifest energy is detectable in such experiments as proposed by the Dutch physicist Hendrik Casimir. Briefly, to simplify, some of the activity of the quantum vacuum is suppressed and with it some of the energy. If one were considering virtual photons, the Casimir effect lowers the number of virtual photons between two parallel plates, and a small force of attraction can be measured between the plates. However, these virtual photons are not promoted into permanent reality without the availability of energy feeding into the Casimir plate system.

out its being a rigid body. The upshot of all this is that we cannot picture an electron in classical terms. Similarly, the energy density of a vacuum is measured as zero but computed to be 10^{94} grams per cubic centimeter. So while we state that the wave function has no manifest energy and space energy density is zero, this may be like the flatlanders saying there is no energy in the vertical direction.

One way to account for the surprisingly high energy level of the calculated space density is to accept the concept of space being tightly packed black holes of the Planck length in radius. Wheeler, in considering space-time as a labyrinth of holes and tunnels, began thinking that maybe a particle such as an electron was like a hole in space-time. If they were black holes, such attributes as mass and perhaps spin could be applied to them. But what about electric charge? Well, Wheeler envisioned a black hole as connected to a white hole in another region of space, but with the connection outside space-time. If an electric field entered the black hole, it would appear as a negative charge, and it would then reemerge from the white hole as a positive charge. This conception explains why electrons and positrons are created together. Wheeler called the whole entity a "geon"[3] and proposed that it might be the fundamental particle of the universe. This fascinating idea of Wheeler's has not been completely developed, but it is a possible steppingstone to other ways of looking at our world.

In 1974 Hawking was investigating the behavior of matter in the vicinity of black holes when he encountered an amazing phenomenon: black holes seem to emit particles at a steady rate. This was contrary to the supposedly established fact that black holes could not emit anything. But the quantum-mechanical concept of tunneling offered a way for particles to escape. The barrier to the emission

---

[3]Wheeler's geons are not particles as defined by the physicist. That is, while they exert gravitational attractions similar to particles, they do not contain "realness" but are made of curved space.

of a particle is inversely proportional to the size of the black hole; if a black hole is large, the emission is small. On the other hand, small black holes emit particles rather rapidly. When the black hole is the size of a quantum black hole, it will literally evaporate in approximately the Planck time. So, matter collapses into a black hole and is lost, but new matter is created in its place, and unlike Wheeler's conception, a black hole is within a white hole. In this process, a black hole loses its charge as it evaporates. In addition, the emitted particles carry away their angular momentum. This concept of evaporating and radiating black holes as the constituents of space will be treated in more detail in Chapter 15.

Motion as projection and injection

We have seen that motion, in the light of quantum theory and the wave function, is quite different from motion described in classical terms. In the Newtonian view, motion is the movement of a localized object from one point to another through space-time. In quantum theory, motion is projection and injection from the primary order of the wave function or, as Bohm would say, an unfolding and enfolding from the implicate order, or level two. This conception is similar in some respects to the movement of a water wave.

According to quantum theory, any two particles (say, electrons) are exactly alike. If that is so, any object in the universe (including us) must be distinguished by its pattern, not by the content of its parts. That is, one could exchange electrons between two bodies, and the two bodies would be the same as before the exchange took place. This brings up the interesting idea of teleporting, which could be used to explain quantum jumps. That is, when an electron goes from one orbit to another, it does not travel in between as a particle. The probability wave does, but not through our three-dimensional world, since the wave function is in level two.

In *The Emperor's New Mind*, Penrose examines the idea of

teleportation as described in a number of science fiction books. If an astronaut wished to travel from earth to a star twenty-five light years away, the journey by a conventional rocket would exceed the lifetime of its occupant since a material object cannot travel at the speed of light. But some science fiction writers circumvent this restriction by imagining that we are advanced enough to determine the exact location of every particle in the body of any person who wished to visit a distant star or planet. This knowledge is recorded in some device, converted to an electromagnetic signal, and sent off at the speed of light. Twenty-five years later it arrives at its destination where the astronaut is recreated. A serious problem is immediately evident: if the initial person is not destroyed, then we have two astronauts with identical awareness — clearly not advisable.

Quantum motion gets around the problem of two astronauts because the electron in one orbit no longer exists when it appears in the second orbit. Nevertheless, the electron in the second orbit is identical to the first one. This is like saying that when you move across the room, you are teleported in the sense that all your constituents are being explicated and then implicated as you move. Another comparison is the apparent movement along a string of lights: if the lights are switched on and off in sequence rapidly enough, it appears that the light moves through space-time. Perhaps movement in general, and quantum jumping in particular, is actually teleportation, with these differences: the information is sent through the second level, so that each constituent is implicated and explicated by information in the implicate order.

Multidimensionality

We have noted that the wave function is multidimensional. But we do not reach an understanding of the wave function domain by adding spatial dimensions to our everyday three-dimensional world. Particles are patterns in Hilbert space, and Hilbert space is not something particles move about in but a domain that houses all the probabilities of action for a quantum system. Each axis represents a probability for a certain action to occur.

Through a device such as Hilbert space, we are able to arrange events or quantum states and see that they are all interwoven and part of each other. In some sense, the path not taken influences the path taken. In the two-slit experiment, the wave function describing the electron combines the electron going through one slit *and* the other one. In some way, the wave function in Hilbert space is telling us that the electron can be in two places at once. The following metaphor demonstrates this. If you roll a die, the uppermost face can turn out to be any number from one through six. A probability is assigned to each face because we agree that the uppermost one is the only one that counts; that is, we define our reality as the uppermost face. Presumably though, a higher-dimensional entity could see and encompass all six faces at once. In other words, more dimensions mean more information and a more developed context.

Could it be that mystics are tapping into this higher-dimensional domain of the wave function and therefore are seeing more faces of the die? Dimensionality may be a measure of the restrictions on our reality. Perhaps we can learn — and are learning — to handle more faces, more dimensions, and therefore more information. And perhaps God, All That Is, can handle a die with an infinite number of faces.

Time and timelessness

Finally, we have found that Schrödinger's equation is time-symmetric, but the act of measurement is not. That is, Schrödinger's equation is deterministic and is in some ways similar to Newton's equations of motion. So if we have a value for the wave function at a given instant of time, we can determine in principle its value into the past or future. But the wave function represents a superposition of possible states, and it is this conglomeration of states that is time-symmetric. So, when we are in the domain of probabilities, quantum theory is time-reversible. But as soon as the real world is created through a measurement, quantum theory requires an arrow of time. When we are looking, we have an arrow of time; when we are not looking, there is no direction to time.

Summary

Our universe originates in the nonphysical hidden domain of the wave function. Each event in our real world has its birth from a sea of infinite possibilities, located in level two — a mysterious realm, a creative source, that interpenetrates our universe. Our analysis of the wave function leads us to the following conclusions about this hidden domain.

- The hidden domain is a vast sea of energy. Though this energy is not detectable by our physical senses or our instruments (not surprising since our mathematical description of it, the wave function, is a complex function), the formalisms of quantum theory tell us that it is enormous.
- The hidden domain contains all possible states for action in our three-dimensional world. Our world of actuality floats on a sea of potentiality or probabilities.
- The hidden domain is described by a *linear* superposition of states, and because of this, an outside element is required for the selection of one potentiality to become actual on our plane. Both Seth and such physicists as Dyson, Charon, and Bohm label that element as consciousness. But consciousness is not restricted to sentient beings. According to Seth as well as some scientists, all energy contains consciousness.
- The hidden domain is one whole and is thoroughly interpenetrated, which means that the world of the wave function cannot be viewed as made up of discrete parts. Phase entanglement tells us that all of reality — including our three-dimensional world — is instantaneously connected, not through a field in space-time, but through the wholeness of the hidden domain.

- ☐ The possibility of faster-than-light particles (tachyons), though proposed by physicists in the early part of this century, has been ignored by some physicists because tachyons have an imaginary rest mass. However, as we have seen in our study, the complex plane and imaginary numbers may well be referring to the hidden domain.
- ☐ Matter is the form taken by consciousness when a probability selected from the hidden domain is projected into the level of space-time. Thus, mind and matter are aspects of a single reality.
- ☐ The hidden domain provides a source from which motion is seen as projection and injection, a view counter to the classical notion which envisioned motion as an entity moving through space from one point to another. A metaphor would be the frames of a motion picture film projected onto a screen in rapid succession, giving the appearance of motion.
- ☐ The hidden domain is multidimensional. The plane on which we live is like a single slice through a higher-dimensional sphere.
- ☐ The hidden domain is timeless. Time in our three-dimensional world is a result of the collapse of the wave function. Just as our three-dimensional space is projected from a higher-dimensional source, time is a projection from a timeless order.

While some physicists view the wave function as merely a mathematical expression useful only for predicting the results of experiments, it is our thesis that it has immense significance and meaning for our conception of reality. The hidden domain, whose qualities we have just related and whose characteristics have such a strong relationship to experiences in our three-dimensional world, is described by the wave function. As this mysterious realm continues to be investigated, perhaps we shall be confirmed in our belief that it is not only part of our reality but is the creative source for our three-dimensional world.

PART 3: MODELS OF REALITY

One of the most impressive aspects of classical physics is that so much about how the universe operates is explained by so few mathematical formulas. As we discussed in the section on the classical view, Newton formulated three laws of mechanics that purported to explain all motion, whether here on the earth or in the farthest reaches of the universe. But what does it mean to *explain*?

This question divides the physics community (although, physicists tend not to be totally in one camp or the other). One group feels that the mathematical explanation of a phenomenon is the final truth and that our knowledge can be considered complete if the mathematics is correct. An example of that view is this statement by Paul Dirac:

> The only object of theoretical physics is to calculate results that can be compared with experiment, and it is quite unnecessary that any satisfying description of the whole course of the phenomena should be given. (Hoffmann, p. 154)

The other group feels, as did Einstein, that a theory should go beyond the mathematical facts and reflect the fullness of reality — it should offer some kind of model, a schematic description or image that enlarges our understanding. The first group, then, regards the business of physics as that of developing algorithms, recipes that order the facts in a consistent way and thereby allow some degree of prediction. This group recognizes the practicality and utility of quantum theory, but feels there is no hope of its resulting in an objective model of reality. Bohr endorsed this point of view, which Bell called FAPP: "for all practical purposes" the procedures are perfectly adequate. The second group holds the view that a good mathematical theory will provide not only appropriate algorithms but the basis for making models of reality.

With the advent of relativity theory and especially quantum theory, modeling reality became much more difficult because the images that arise from the mathematical formalisms are extremely counterintuitive, that is, not at all what we are accustomed to in our everyday sensory experience. Indeed, the whole concept of an objective reality has come into question. It would seem from the Copenhagen interpretation, and certainly according to von Neumann's view, that some sort of observer is necessary to create reality. Perhaps a better way of saying it is that "something" is needed outside the quantum system. What that something is, is a matter of considerable conjecture. Since all measuring systems are themselves quantum systems, we know the something must be outside of space-time, or outside our three-dimensional world.

Since the quantum state is described by the wave function, a model for the wave function should be helpful, even if the result is counterintuitive. Perhaps we can learn from our effort to develop such a model, difficult as that process may be. Models have already been set forth by some physicists. In considering a number of these, it soon becomes apparent that each model appears to be touching a different part, as it were, of the "elephant" of reality. But by incorporating aspects of each of these views into a broader, more encompassing construct, we may develop a clearer and more complete picture.

14 Quantum wave modeling

The official view

To sum up Bohr's view in one sentence, we might say, "There is no real quantum world; there is only a mathematical description." Bohr did not say that the world we perceive is an illusion but rather that the world of our senses floats in a domain that is not real and objective. If that is so, quantum theory is a description of the relationship between Newton's ordinary world of experience and the world of the quantum, which we cannot directly experience. The classical approach starts with elementary particles, atoms, and molecules as the constituents from which reality is constructed. But the Copenhagen interpretation starts with our macro world, and states that the quantum world is created only by observation with a measuring instrument. It is important to note the paradox inherent in the Copenhagen view: if the macro world Bohr starts with is made up of elements from the quantum world, and if the quantum particle is defined by a real-world instrument, then the real world requires the quantum world as its source of constituent elements, and the quantum elements need the real world to bring them into reality. This is a bit like lifting yourself up by your own bootstraps.

The role of mind

An important distinction needs to be made here regarding the human visual system, which is itself a measuring instrument, and the mind. It is the mind — not the brain — that collapses the wave function. Consciousness, an attribute of mind, seems integral to quantum theory, but how can it be addressed in physics? Bohr left that unclear, to say the least.

In attempting to determine the exact point at which the wave function collapses to bring the quantum world in reality, von Neumann concluded that it could be anywhere between the original quantum state and the observer's consciousness. Perhaps a better way of stating this is that if one accepts the mathematical formalism of quantum theory, then there is no dividing point marking the collapse of the wave function, no natural break between the quantum event and the mental observational experience, except at the level of the mind. The reason, of course, for von Neumann to settle on the mind is that it is the only element outside the system. In short, logic and the mathematics of quantum theory tell us that the Schrödinger equation is applicable throughout the physical world, including the macroscopic level, in spite of our conditioned belief that there is a world "out there" independent of the observer.

In his book *Speakable and Unspeakable In Quantum Mechanics* Bell describes various interpretations of quantum theory without necessarily endorsing one over the other. This is the way he describes the concept that the existence of the material world depends on the participation of the mind:

> It accepts that the "linear" wave mechanics does not apply to the whole world. It accepts that there is a division, whether sharp or smooth, between "linear" and "nonlinear," between "quantum" and "classical." But instead of putting this division somewhere between small and big, it puts it between "matter" (so to speak) and "mind." When we try to complete as far as possible the quantum theoretic account of the electron gun, we include first the scintillation screen, and then the photographic film, and then the developing chemicals, and then the eye of the experimenter . . . and then (why not) her brain. For the brain is made of atoms, of electrons and nuclei, and so why should we hesitate to apply wave mechanics . . . at least if we were smart enough to do the calculations for such a complicated assembly of atoms?

> But beyond the brain is . . . the mind. Surely the mind is not material? Surely here at last we come to something which is distinctly different from the glass screen, and the gelatine film . . . . (Bell, p. 191)

Levels of reality

Another extension of the Copenhagen interpretation is Heisenberg's division of reality into two parts, the everyday level of actuality and an underlying level of potential, which is described by the quantum wave function. We have seen that the wave function contains all possible information relating to the quantum system and that the properties of the system do not exist separately and in a precisely defined form but only as incompletely defined potentialities. So, according to Heisenberg, the world of the wave function is constructed of unactualized potential. Also we have seen that the wave function evolves according to the Schrödinger equation in a continuous and deterministic way with respect to time. This process, however, does not control the evolution of actual events but merely describes the tendencies for actual events to occur. Events consist of quantum jumps which are not *individually* controlled by any known law of physics. In aggregate they conform to statistical rules, but each particular event seems to be random. So, Heisenberg is as vague as Bohr on how this potentia is brought into our world, though he agrees with Bohr, of course, that an act of measurement is necessary.

Parallel universes

The mind as the location of the wave function collapse is unacceptable to many in the physics community. Einstein himself asked incredulously, "Can the mind of a mouse collapse a wave function?" In 1957 Hugh Everett suggested a model that would eliminate the collapse altogether by saying something to the effect that if something can happen, it will. Since all systems have an associated wave function, including the measuring instrument, when a quantum system is measured, there are a number of possible systems, each with a measuring device indicating a different outcome. Everett was aware that we never experience measuring instruments split up into alternative possibilities, yet he wanted to get rid of the offending concept of collapse. If there are all these possible systems and yet we see only one of them, what happens to the rest? One way to get rid of the ones we do not see is to consign them to parallel universes. That is, Everett says that upon each measurement of a system that has alternative possibilities available to it, all *possible* outcomes actually do occur, but this happens in new universes, all of which are exactly the same *except* for the alternative possibilities under measurement. The result: no wave function collapse.[1]

If Everett is correct, we may ask why humans beings live in only one of these universes at a time. Everett explains as follows. Suppose there is a 50% chance that an electron can go up and a 50% chance that it can go down. We then have two parallel universes, one with the electron going up and a second with it going down. We also have an observer in each universe. According to Everett, our bodies are part of the universe just as an electron or a distant star is, and copies of ourselves are constantly splitting and shooting off into those parallel universes. These copies are not mere automatons; they think and feel just as we do, and each inhabits a universe similar to our own. Since we are not consciously aware of these replicas of ourselves, Everett and his followers believe that these separate worlds are completely disconnected and isolated.

Everett's model is called "the many-universes interpretation of quantum theory,"[2] and the conglomeration of worlds it describes

---

[1]Recall that in Hilbert space, the wave function is represented by a vector that is actually the sum of any number of vectors at right angles to each other. Everett says that each of these vectors represents a parallel universe.

[2]Also called "branching universe" to compare the many universes to the many branches of a tree. The tree is analogous to the superreality of all universes while we live out our lives on a single branch.

is called superspace. In this many-universes view, our three-dimensional world is just a subspace of the overall superspace. We cannot visualize superspace, but it can be described mathematically. The idea of vast numbers of parallel worlds (perhaps 10^{100}) might seem more like fantasy than fact, yet, it has a substantial following among theoretical physicists who are interested in modeling formalisms of quantum theory.

The participatory universe

While Everett's many-universes concept succeeded in ridding physics of the collapsing wave function, it does not eliminate the need for an observer. Each branch of the many universes still requires a measurement at the point of bifurcation (the point at which each universe splits off), which means we are back to square one since measurement implies an observer. A few physicists have taken the position that the branching is real enough since it is described by mathematics, but that it is consciousness that decides when branching occurs.

Some physicists use Heisenberg's potentia to describe the many universes; that is, the other universes created by incessant branching exist in potential form but are not real in our sense. This idea corresponds somewhat to that of a "participatory universe" developed by Wheeler, who has given a tremendous amount of thought to the meaning of quantum theory.

An analogy for the participatory universe is given in Wheeler's description of playing "Twenty Questions" with a group of friends. The idea of the game is that one person is sent from the room, and the rest decide on an object for the person to try to identify by asking questions of the group. When Wheeler was sent from the room, the group decided not to agree on an object. Instead, each participant answered Wheeler's questions truthfully about an object in his or her mind. Wheeler finally guessed the answer to be a cloud, but of course there was no agreed-upon object. Through his questioning, Wheeler himself, in effect, created the cloud. This version of the game, Wheeler says, is similar to quantum theory in that an electron is "created" by experiments, which are, in Wheeler's terms, the questions physicists put to nature. He sums it up this way:

> No elementary phenomenon is a phenomenon until it is a registered (observed) phenomenon. . . . Useful as it is under everyday circumstances to say that the world exists "out there" independent of us, that view can no longer be upheld. (Wheeler and Zurek, pp. 192, 194)

To Wheeler, a fundamental element of quantum theory is the act of measurement, which implies participation by an observer, hence his belief in a participatory universe. Here is how he states it:

> The choice one makes about what he observes makes an irretrievable difference in what he finds. The observer is elevated from "observer" to "participator." What philosophy suggested in times past . . . quantum mechanics tells us today with impressive force: In some strange sense this is a participatory universe. (Wheeler, p. 25)

One example of how this works is demonstrated by the two-slit experiment described earlier. Recall that putting a detector at one slit in an attempt to discover which slit an electron went through destroyed the interference pattern. If no attempt was made to discover the route of an individual electron, then the interference pattern reappeared. So in setting up the two-slit experiment, an interference pattern can be produced or not, depending on the experimental arrangement, or on what question is asked.

Wheeler came up with an ingenious thought experiment that revealed a tremendous insight. In his "delayed choice" version of the two-slit experiment he supposes a way of switching rapidly from one experimental arrangement to another. If the switching of experiments is fast enough that the decision of which experiment to perform can be made *after* a particular electron goes through the slit, then the physicist's decision could influence the electron in motion by reaching into the past. That is not to say one could send a signal into the past; that is not possible because in quantum theory action is unpredictable. Yet in some sense, in this experiment the physicist creates the past in the present.

You might think that since this is a thought experiment, the anomaly can be disregarded. But an actual experiment testing the hypothesis was performed by Carol Alley and co-workers at the University of Maryland using laser light striking a half-silvered mirror, which divided it into two beams analogous to the two-slit paths. Sure enough, the path the photon actually took was determined after one of both paths was already chosen, thus validating the philosophical idea that the past has no existence except as it is recorded in the present. Basically, this experiment indicates that Newton's view of time as a sequence of events from past to future is no longer tenable. That is to say, even time is nonlocal in that the past, present, and future are completely interpenetrated.

It is tempting to think that while delayed choice has validity on the particle level, it cannot be applied in the macroscopic world. But everything we see is made up of particles that obey these rules; therefore, everything that is real in our physical world is built of elements that are not real until we make them so. In some way or another, everything real is created by an astronomically large number of acts of participation by an observer or observers. Thus we could say that when a physicist observes the photons of the background radiation of the big bang, he or she is creating the universe — that in fact the universe existed only in potential form until the observers came along and made it real.

Wheeler's ideas demonstrate that quantum measurement is essentially a downwardly causal activity. Classical physics, remember, is based on upward causation, with events moving forward in time: an atom gives off a particle, the particle engages a piece of experimental equipment, a pointer on a meter moves, and a physicist then reads the meter. With Wheeler's delayed choice experiment, however, there is a closed circuit, beginning with retro-causation, which goes back in time, and ending with forward movement. The process starts with an idea in the physicist's mind: to measure an electron emitted from an atom. The electron exists in a state of potential until the physicist's measurement creates it. This is the downward leg; it moves backward in time. The electron then engages the equipment and deflects the needle, which registers on the physicist's optical system and mind, completing the circuit with movement forward in time. In a sense, then, the process begins with mind and ends with mind.

A handshake in time

Wheeler says that the observation, or measurement, is the starting point in quantum theory and the electron is a product of this primary activity. Thus, "elementary particle" is a misnomer, since particles are secondary, being derived (by observation) from a more fundamental level. This circular movement, starting with downward causation, then completing the process with upward causation, is similar to Cramer's transactional interpretation, discussed on pages 73–74. Cramer's idea is that reality is created by the wave function going backward *and* forward in time.[3]

As we have noted, a quantum wave function describes a group of possibilities, any one of which a quantum system could assume. The wave function can be imagined (although not measured) as a wave spreading throughout space with each point on the

---

[3]Paul Davies uses the analogy of the particle as computer hardware and the wave function as software. He sees the behavior of the particle as influencing the form of the wave function and, in turn, when a measurement is made, the wave function influences the future behavior of the particle. Since they are entangled, Davies sees

wave representing a position possibility for the entire quantum system, with one possibility becoming fact when an observation takes place. No one seems to know exactly how or where the wave function collapse occurs. We have a clue, however, when we remember that to get a real number probability we multiply the original waves of possibility by their complex conjugate wave. As we have seen, Wolf interprets Cramer's transactional approach in an interesting way. If for a moment we consider the quantum wave as real, then the complex conjugate wave is also real, but it travels back in time, in the opposite direction from the original wave. When the two meet up, a "real" probability is created. This concept is similar to the Wheeler-Feynman theory which postulates that any transaction requires an advanced and a retarded wave. The advanced wave is exactly out of phase with the retarded wave and the two cancel. We then assume that the original retarded wave is absorbed and the advanced wave is thereby not detected. Cramer's approach is actually quite complicated, but the essential point is that an event comes into existence through the interaction of two waves. One wave travels forward in time and is quite familiar to the physicist; the second travels back in time and is undetectable. So if for the sake of argument we use Wolf's notion that the wave function is analogous to a real retarded wave, then the complex conjugate is analogous to an advanced wave. We shall examine this method of creating events later in Chapter 15.

Unfolding and enfolding

Let us conclude our discussion of quantum wave modeling with the remarkable concepts of Bohm. Bohm's basic thesis is that

---

the measurement problem as the result of artificially trying to untangle them. If we look at the measurement problem in this way, the particle activity gives rise to effects registered by the measuring equipment, the observer looks at the equipment, and then (through Wheeler's delayed choice) the observer couples back into the particle activity. This creates a closed loop comparable to the process Cramer describes.

reality is multilayered. Our three-dimensional world, the level he calls the explicate order, contains matter, space, and time. Bohm postulates another order, which he calls the implicate order — a hidden level that permeates our three-dimensional world, an enfolded domain underlying the explicate order. Physics operates largely in the explicate order, but its equations extend to more fundamental levels. While the utility of these equations is clearly understood, their meaning is not.

Bohm tells of watching a television program in the early 1960s in which a fascinating object was shown: a glass jar with a rotating cylinder inside, the cylinder turned by a handle on top, and between the jar and cylinder a narrow space filled with glycerin. The glycerin was light-colored and viscous. A drop of ink was put in the glycerin, and the handle was slowly turned. The ink began to smear away into nothing, seemingly absorbed into the liquid. Then the rotation was reversed. Amazingly, the original drop of ink reappeared. That process provided the image Bohm was looking for: the ink drop represented the unfolded or explicate order while the glycerin with the ink smear represented the enfolded or implicate order. This image shows the universe as one whole, with forms we see explicated from the whole then folded back in again in a ceaseless succession of enfolding and unfolding.

The whole enfolded order cannot be made manifest to us. Only some aspects are unfolded, giving us the experience we call perception. The classical view assumes that the whole can be unfolded, even though instruments may be required to see it in its entirety. Also in the classical approach, motion involves an object crossing space over time, but in Bohm's vision, motion is the expression of enfolding and unfolding. This leads us to conclude that connections in the enfolded order are beyond space and time, a conclusion that helps explain the apparent conflict between Bell's theorem and the special theory of relativity. That is, relativity applies to the explicate order, not the implicate. On the other hand, the existence of the implicate order is consistent with the nonlocality requirement of Bell and the instantaneous connections of quantum events. The invisible, nonlocal reality that Bell requires is the enfolded level, or implicate order, that Bohm postulates.

As we discussed earlier, the concept of zero-point energy, or what we call empty space, is really a vast ocean of energy. Physicists ignore this energy because it is not measurable. Using the analogy of the ink and the glycerin, it is as if the physicist can measure only the energy in the ink drop when it is visible but not when it is immersed in the glycerin. So in some sense, the ink drop, analogous to the physical reality of the explicate order, is like a small ripple that appears on the surface of the vast ocean of energy of the implicate order.

In addition to the implicate order, Bohm postulates a second enfolded order — a super-information field of the whole universe. Recall that the wave function (or the implicate order it describes) is linear or passive in that it requires something to organize it and to create relatively stable forms from it. That organizing function is provided by what Bohm refers to as the superimplicate order. Bohm assumes that there is a continuum of orders, which merge into an infinite-dimensional ground he labels the holomovement.[4] It is important to emphasize that Bohm's concept of the implicate order, though perhaps not as widely understood, provides an interpretation that is just as compatible with the equations of quantum theory as the well-respected Copenhagen interpretation and, for that matter, any other explanations of that stature.

Bohm's concept is a new and intriguing way of viewing the relationship between the material and the mental. The idea of an

[4]See pp. 63-73 in *Bridging Science and Spirit* for more detail.

unmanifest aspect of reality (implicate order) and a manifest aspect (explicate order) is analogous to the relationship between consciousness and matter. As Bohm says:

> There is a similarity between thought and matter. All matter, including ourselves, is determined by "information." "Information" is what determines space and time. (Zukav, p. 327)

This insight led Bohm to what he calls soma-significance. Unlike the term "psychosomatic," which implies two entities — mind and matter — in some sort of interaction, soma and significance are two aspects of a single process.

Each somatic form has a particular kind of significance associated with it; for example, a printed page is soma, which has significance when apprehended by a reader. The significance is carried by physical processes in the brain and understood by the mind. Thus, what is soma on one level can be significance on another, and what is significance on one plane can be soma in a broader context. Reality, then, is constituted of subtle levels displaying different aspects. In Bohm's view, even the electron has a protointelligence, which is subtle compared to its more material aspect. Finally, since all activity proceeds through enfoldment and unfoldment, there seems to be a two-way movement of energy, with each level of significance acting on the next somatic level and perception carrying the action back in the other direction.

In *Reality and Empathy* Alex Comfort employs a metaphor which nicely illustrates levels of order which we have concluded are necessary to model quantum reality and, more specifically, the wave function. He asks us to imagine a computer game consisting of a black box containing the electronic circuitry and a screen where the action is displayed. At this point, we have a two-level analogy, with the screen similar to the explicate order and the black box representing the implicate order. A third element is a control for the player who by looking at the screen can cause or prevent collisions — the point of the game. According to Comfort, the player represents Bohm's superimplicate order. So the function of the display (explicate order) is to produce a manifest world that can develop the skills of the player. The black box of electronic circuitry (implicate order) contains in a latent or potential form all possible events that can be displayed on the screen. If the display on the screen develops the player, can it be that the explicate order of our everyday world provides a means for our own evolution? That captivating question leads us to an examination of Seth's views.

15 Modeling with Seth

In this chapter, we turn to the stimulating ideas expressed by a discarnate entity called Seth who was channeled by the late Jane Roberts. In September 1963, Roberts had what she described as her first paranormal experience. While writing poetry, she suddenly felt her consciousness leave her body, and at that moment she began recording a torrent of information that flooded her mind. These writings are the enormously provocative definitions entitled "The Physical Universe as Idea Construction."[1] An amazing aspect of this encounter for Roberts was that not only were some of these ideas new to her, but many were contrary to her own beliefs at the time. Later she realized that "The Physical Universe as Idea Construction" was the embryo from which all the Seth material eventually evolved.

At the time she first encountered Seth, Roberts was researching and writing a book on extrasensory perception, the acquisition of information outside the five senses. As part of her project, she and her husband Robert Butts began experimenting with an ouija board. After a few sessions, they began receiving messages from someone or something called Seth — an "energy essence personality" (Seth's term). After a few more sessions, Jane began anticipating the ouija board messages, speaking them before the pointer moved.

---

[1]In *Seth: Dreams and Projection of Consciousness*, Roberts described her experience in the following way:

> Exultation and comprehension, new ideas, sensations, novel groupings of images and words rushed through me so quickly there was no time to call out. There was no present, past or future: I knew this, suddenly, irrevocably. (pp. 42-43)

When her consciousness settled back in her body, she found that about three hours of clocktime had passed. She also discovered a pile of scribbled notes, forty pages in all, describing the images and ideas she had encountered. Among other things, there were definitions for energy, ideas, instinct, and physical time. For interested readers, a large portion of the original manuscript is given on pp. 44-51.

Soon after that she began to enter a full trance in which she allowed Seth to use her body as a vehicle to deliver his message. These sessions resulted in many books written by Roberts herself as well as those dictated by Seth. Of the 1,788 sessions, all but 252 were transcribed into written records. Much of this material has been published, including ten books dictated directly by Seth.

In this chapter we will find that Seth's view of reality conforms to portions of many of the interpretations of the quantum world that we have considered, reflecting facets of various physicists' ideas. That alone makes the Seth material fascinating to me.[2] In a similar way, John Gribbin considered the range of scientific models of the quantum world and "bought the lot;" in his book *Schrödinger's Kittens and the Search for Reality* he says:

> [As far] the question of which of the many quantum interpretations, if any, can be regarded as a best buy. . . . The best answer, it seems to me, might well be to buy the lot. Each of the interpretations is a viable model, and each of them provides us with useful insights into the way the world works. Indeed, it is quite reasonable to regard each of the quantum interpretations as a degree of freedom in its own right . . . we are free to choose the interpretation that most suits our needs in any particular situation. (Gribbin, p. 219)

With that in mind, we will examine Seth's concepts in relation to those discussed so far in our attempt to reach a fuller understanding of reality.

Frameworks

One of Seth's most interesting ideas is frameworks of existence. We have encountered the proposal that reality exists on at least three levels: ordinary three-dimensional space-time, the underlying domain of potential, and the realm in which potential is

[2]Some of Seth's concepts are covered in greater detail in *Bridging Science and Spirit*.

brought into actuality (see pp. 147–151 for Bohm's formulation of this). Seth's views are in basic agreement with Bohm's idea of the implicate order. Bohm sees an endless series of implicate orders, comparable to Seth's frameworks. In Bohm's conception, all these orders are aspects of the holomovement, or the ground from which the universe is manifested. Seth says in an analogous way that these frameworks are aspects of All That Is. Both see space and time as projections from this ground and the ground as totally interpenetrated; rather than localized entities moving from one place to the other, the ground contains All That Is at every hypothetical point. (A hologram might be seen as a three-dimensional metaphor for the ground.) Dividing All That Is into levels is only a map we use in our attempt to understand — the ground is actually totally interconnected and interpenetrated.

The similarities between Bohm's and Seth's descriptions of the three levels are striking. Both agree that the three-dimensional world we live in (the explicate order, or Framework 1) is unfolded from a second order (the implicate order, or Framework 2). Both also agree that a third level (the superimplicate order, or Framework 3) is necessary to make selections from the second order. That is, the third level houses the choreographer or conductor of our world, the agent who selects from the probabilities of the second level to create the first level. (Seth says when one dies, the part that survives returns to the third level; Framework 3 is where Seth says he encountered Roberts's consciousness.)

The concept of frameworks, or orders, as Bohm would phrase it, is almost mandatory if Bell's theorem is taken seriously — and this theorem must indeed be taken seriously since it is backed up by experimental proof. According to Bell's theorem, reality must be nonlocal. However, special relativity theory, which has been with us since 1905 and has never failed to give correct answers, absolutely rules out the possibility that the universe is nonlocal. How do we reconcile this immovable object and this irresistible force? Seth, Bohm, and a few others do it with the concept of levels, suggesting that relativity theory is remarkably accurate in Framework 1, but we need the nonlocal quality of Framework 2 to support local phenomena in Framework 1.

Framework 2 often appears to be completely hidden from our everyday perceptions. This may be because we have not learned to adequately understand such experiences as extrasensory perception and altered states of consciousness. Perhaps the physics community has reduced the sphere of its concerns too drastically; it has ignored and even ridiculed areas of investigation that might give important clues regarding this mysterious underlying nonlocal realm. If these areas were addressed with open minds and with careful scrutiny, then we might begin to understand that not only is a hidden domain necessary but also that it is the source of our three-dimensional world.

Consciousness as energy

Seth also is in agreement with certain aspects of the Copenhagen interpretation, especially what might be called Heisenberg's understanding that the quantum world cannot be defined by objective events in time and space but is in a state of potentia, expressed as probability waves, and comes into being only by an act of observation. Wheeler's participatory universe also has elements that are congruent with some of Seth's ideas. Recall that Wheeler assumes that the physical world is actualized through an act of observation. That is certainly compatible with quantum theory but involves a paradox: the physical world that is actualized presumably creates the observers that do the actualizing. Wheeler, of course, was aware of this problem and suggested that the physical universe in some sense bootstraps itself into existence.

Seth's assertion that energy is the basis of the universe is confirmed by Einstein's special theory of relativity, which states that there is a relationship between mass and energy mediated by the velocity of light squared. That is, mass is condensed energy; thus the entire universe is basically energy that can be condensed into what we call matter. But Seth goes one step further. He emphatically states that all energy contains consciousness, that everything in the universe is conscious at some level. In our discussion of the wave function and consciousness, we considered the idea that the electron "chooses of its own free will" to jump from one orbit to another, or, in Dyson's words, "an electron is an active agent making choices." With quantum theory, the electron is seen to be both a particle and wave, going from one state to another without passing in between — a phenomenon that can be described mathematically but which is difficult to understand conceptually. To help make sense of this, Bohm postulated a minimal protointelligence in the particle. As we have seen, all this arose because of the necessary role of observation in quantum theory. Observation implies consciousness, and given the concept of phase entanglement and the resulting interconnectedness, then if there is consciousness anywhere, it must be everywhere. What Heisenberg calls potentia, Seth calls thoughts which have not been explicated into matter. And what Bohr calls an observation, Seth calls an act of creation by the observer. That is, according to Seth, the thing perceived is an extension of the one who perceives it.

For the moment, let us accept the equivalence of energy and consciousness. Seth describes basic units of consciousness (CUs) as being analogous to particles but without physical substance. Since energy is not divided, he says there can be no separate portions of it, so the terms "particle" and "basic unit" are misleading. A better way to grasp this idea is to think of a CU as a ripple in a vast ocean of energy. In one sense, it can be seen as a separate entity, but in a broader context it cannot be separated from the ocean. We shall employ the CU terminology because it helps convey some complex ideas. For example, this small unit of energy does not have weight or mass in our three-dimensional world but does have the potential for the creation of matter in all of its forms and the impetus to create all possible universes. Thus, if we think of all reality as All That Is, and equate that with God, then CUs are fragments of the divine.

In Jungian terms, the pure energy of the CU can be described as having the creative propensity toward individuation,[3] which is the developmental process toward wholeness or integration. A metaphor for this process is a plant that from its bulb unfolds its latent abilities fully, developing naturally into whatever it was meant to be. Since we have seen that wholeness is basic both to physics and to Seth, individuation is a drive to broaden the context of each consciousness. All conglomerations or gestalts of CUs — all entities — have the ability to transform energy into an idea and then to give the idea physical form; a reflective mind is not required. With amoebas, for example, the few instinctual ideas received by them are constructed almost instantaneously since these organisms are not far enough along the evolutionary spectrum to reflect upon these ideas. But as one goes up the scale, memory comes into play. That is, the organism begins to have built-in images of past constructions which are used to test and perfect new ones. Choice appears; though the reflection period is brief, an animal can choose, for example, to attack or not to attack. With humans, it becomes impossible for the organism to construct all its ideas physically. Therefore, choice becomes necessary, and self-consciousness and reason emerge. Time becomes

---

[3]Individuation denotes the process by which a person becomes whole or comes to self-realization. Such a process does not shut one out from the world, but rather it gathers the world to oneself. In that sense, CUs seek out their individuation by creating matter in all its forms, including creating multiple universes.

important, for the human organism is no longer fixed in the present. Also, in humans, memory does not flicker so briefly as in animals, but stretches out behind and affects what lies ahead in the form of influencing future choices. Through this process the human ego was born, with its division of experience into subject and object. In some mysterious way our consciousness is both the construction engineer and the construction itself.

According to Seth, CUs can operate as particles or waves. When CUs operate as particles in our universe, they build up a continuity in time and enter space-time (Framework 1). Recall that the Schrödinger equation is time-symmetric. That is, while the electron is still defined by the wave function, whether the quantum state is going into the past or the future is irrelevant. However, when the probability equation is applied and a measurement made, an arrow of time is created. This is what Seth means by "building up a continuity in time." In the two-slit experiment, the electron showed its particularity — a definite mark on the photographic film — when it changed from a wave to a particle. So Seth's description of CUs as both waves and particles bears a close relationship to the wave function description of the electron.

Not only are CUs the building blocks for all the material of the universe, they are also responsible for the manifestation of so-called empty space as we know it. How are we to understand this Sethian view? In Wheeler's conception of space-time, we found that space has a tremendous energy density — calculated (not measured) to be 10^{94} grams/cm^3, much greater than nuclear density, which is in the neighborhood of 10^{14} grams/cm^3. Such a comparison seems preposterous in normal terms. If, however, we consider "empty space" as not being in the realm of space-time but as the underlying domain of potential (Framework 2), and think of matter as the realm of manifestation (Framework 1), then such a comparison makes sense. One way to identify these levels is by their accessibility to physical instruments, and the energy density of space cannot be detected by an instrument. Nevertheless, theoretically we can consider space to be the home of a tremendous density of virtual photons, virtual particles, and Wheeler's "wormholes," all of which Seth says are unexplicated CUs. Space, then, is not an emptiness with occasional bits of matter scattered about but rather is alive and teeming with potentiality for creating our universe.

The big bang theory, with its expanding universe and outer galaxies bursting into nowhere, is a misconception, according to Seth. Instead of resulting from some ultimate explosion, our universe is formed from a deeper reality — the hidden domain, the home of the wave function. This level of reality does not exist in space or time but rather is a psychic domain which forges and shapes our three-dimensional world and other worlds as well. In this view, stars are like icebergs formed out of the medium, water, that gave them birth — except that the medium of energy giving birth to stars is alive with the need to actualize its potential for its own individuation. How are the "icebergs" of matter actually constructed from the ocean of energy? Or, how does a CU go from a wave to a particle? If all gestalts of matter are arrangements of CUs, but CUs have no measurable mass or energy in our universe, how can there be a body that has mass? To understand the answers to these questions, we must turn to the Sethian concept of electromagnetic energy units, which emanate from consciousness in all its forms.

Creation through electromagnetic energy units

An electron is a stable particle in that if it is left to itself, it never breaks down into smaller segments. The same is true for the positron (anti-electron). Since their charges are opposite and cancel each other, when they meet they are mutually annihilated. But since energy is conserved, something is left — a photon with the equivalent energy level. This suggests that particles are condensed electromagnetic energy. On the other hand, we have seen that the photon is the particle of light. So, is a particle condensed electromagnetic energy? Bohm answers this way:

> . . . [M]odern physics thinks of the electron as moving back and forth, approaching the speed of light . . . . In a way, the energy is being reflected. That is, instead of going straight on as with light, it gets reflected somehow back and forth. Now that reflection is what turns energy into matter. You see, the energy is condensed. And if the reflection ceases, the matter turns into energy. But it always was energy. So matter arises when there arises a pattern which reflects back and forth and becomes stable. (Factor, p. 123)

In this view, particles can be seen as condensed light.

We know from relativity theory that the entire electromagnetic spectrum travels at one velocity in a vacuum, 300,000 km/sec. Seth, however, says that our light is only one portion of a large spectrum of velocities. We are not aware of the whole spectrum because our physics deals only with light that is experienced in our three-dimensional world. That is, particles of faster-than-our-light energy are not perceived by us. Remember, though, that Einstein's equations do not prohibit a particle from attaining faster-than-light velocities if such a velocity was there at the moment of its birth (p. 23).

Electromagnetic energy (EE) units are Seth's name for particles that travel faster than light. If these superluminal EE units were to drastically reduce their velocity to conform with that of our universe, then it would be possible for us to perceive them as matter. This is the basis for Seth's concept that consciousness creates matter.

Seth says that the intensity of EE units is determined by the intensity of the thought or emotion of the CUs, that is, of the conscious entity. In this way mental activity has a catalytic effect, rather like enzymes. Just as chlorophyll converts light energy into chemical energy, the mental activity that creates the EE units converts energy into matter. This conversion is accomplished by slowing down the EE units to the velocity of light in our universe, and at that point matter is actualized. If the EE units do not slow down enough, they become latent matter, referred to by physicists as virtual particles. But are we completely unaware of such faster-than-light units? Seth says not always: dream events, for example, are composed of such particles. When we dream, we enter another reality, or another universe, in which we experience images fashioned from particles going faster than light.

Seth says there is a constant surge into our universe of new energy through a near-infinite number of minute sources. That is, Framework 2 constantly pulses energy into Framework 1. The CUs themselves act as extremely small but potent black holes and white holes, through which the energy is sent in the form of EE units. Recall that in the discussion on pages 103–104 we entertained the thought that space can be seen as a near-infinite sea of fluctuating black holes. This proposition poses a difficult problem for physicists for two reasons. First, it throws into question our normal concept of space being a container. That is, with this view space cannot be described by points as in the Cartesian view. Second, the classical laws we know do not apply in a black hole. Nevertheless, this conception certainly resonates with Seth's idea that the CUs are a near-infinite number of minute energy sources for our universe.

It might seem that if our space is filled with black holes, matter and information would only leave our universe, which would hardly be conducive to a stable system. But what Seth describes are pulsating black holes and white holes; the black hole is within the white hole, so that energy is removed from our universe during the black hole phase and returns during the white hole phase. While some physicists envision space as an almost infinite sea of very small black holes, how do white holes come into the picture? Hawking discovered that black holes emit energy at a steady rate, and since a quantum black hole will disappear in approximately the Planck time, Hawking interpreted this as a black hole within a white hole. However, Seth says that not only is a black hole within a white hole, but there is a rapid cycling between the two. Since the charge also changes with each cycle, the black hole may have a positive charge, but on the return cycle it may have a negative charge.

Wheeler's concept of the geon as the basic building block of a particle, you will recall, is a black hole in space connected through a tunnel outside our universe to a white hole opening into another part of our universe, with an opposite charge in the white hole created by an electric field entering the black hole. In that way, Wheeler explained why electrons and positrons are created together. But where are all these positrons in other parts of our universe? Seth may provide us with a clue: perhaps the geon did not tunnel through to other parts of our universe but actually went on to another universe. If Seth is correct in his cycling concept, then our entire universe is made of either matter or antimatter, depending on the portion of the cycle we are in. The periods of this cycle are said to be extremely short. If we assume that Seth's "extremely short" time is equivalent to the Planck time, then we have an alternating matter and antimatter universe which cycles in and out every 5.3×10^{-44} seconds. About matter constructed of different velocities of light, Seth makes the astounding statement:

> These are all variations, generally speaking and very simply put, on matter as you think of it. The same applies to negative or antimatter however, which you do not perceive in any case. (Roberts, *Seth Speaks*, p. 362)

Seth is saying that in the antimatter phase of the cycle, we are not aware of the matter of the universe.

Physicists' current explanation for the absence of a significant amount of antimatter in the universe is as follows. In the beginning, the universe was a gas consisting of all the particles now known, which, with their antiparticles, were constantly created and annihilated. Through this process most of the original particles and their antiparticles were lost. Theorists speculate that there was a slight excess of matter over antimatter, approximately one part in 10^9. When matter is created from energy in a laboratory, there is always an equal creation of matter and antimatter. However, at the temperature of the big bang, it is theoretically possible that electrons, for example, could have outnumbered positrons by one part in a billion. These leftover particles of matter went on to constitute atoms, stars, and eventually you and me. According to this theory, this initial condition must be accepted to explain matter and antimatter if we are still to have a universe. Obviously, this concept does not describe a universe flickering on and off.

Seth's cycling concept may be modeled by Pauli's image of the ring as the complex unit circle (pp. 128–129). As the ring is rotated, the vector goes from the real axis into the imaginary world and returns to the real world as antimatter; another rotation brings the ring back to the real axis. The complex unit circle is consistent with Seth's description of pulsations of EE units in and out of our world through the black/white holes.

If Seth is correct, our universe is flickering on and off every 5.3×10^{-44} seconds, which explains the problem known in physics as "the watched pot never boils." An argument often used to counter the idea that consciousness is the outside agent that collapses the wave function is that one could prevent change in a quantum system by continually staring at it. In this way, the system would not evolve with time according to the Schrödinger equation since the wave function would remain collapsed. But the pulsations described above would preclude such a situation since the wave function and all its probabilities are reestablished every 5.3×10^{-44} seconds. In this way, our reality is more like a motion picture, but with the individual frames going by extremely fast — at the undetectable interval of the Planck time.

There are also interesting similarities between Seth's conception of EE units slowing down to become matter and Cramer's transactional interpretation (pp. 146–148). Recall that the original quantum wave goes forward in time, and the complex conjugate wave goes backward in time, with a probability selected by the two waves meeting. We compared this idea to the Wheeler-Feynman theory that any transaction, to be complete, requires an advanced and a retarded wave, which are exactly out of phase and cancel each other. The original retarded wave is absorbed, the advanced wave goes undetected, and a kind of handshake occurs between them whereby an event is created. In Seth's conception, the EE unit emitted by a consciousness is comparable to the advanced wave. Since the EE unit is going faster than light, it is going backward in time. The retarded wave is the electromagnetic wave which goes forward in time and strikes our eyes. Thus, according to Seth, matter is created.

The modulation of energy through coordinate points

We noted earlier that EE units, which are emanations from CUs, drastically lower their velocity to form matter. According to Seth this is done by the intensity of the thought or emotion which originally caused the radiation of EEs. To understand this we need to return to our description of the superluminal particles called tachyons. Unlike other particles, tachyons can move faster than light since they started out that way. But to slow them down enough to enter our universe requires near-infinite energy. If this energy were available, then the CUs or EEs could create a black hole (see p. 102, footnote 6) and a singularity, the center of a black hole which is infinitely dense and thereby creates a hole in space-time. Now, if this is applicable to tachyons, where does this energy come from? Seth uses the concept of coordinate points to explain.

There are other realities, other universes, Seth says, made of particles that do indeed travel at different light velocities. Normally, consciousness in one reality does not see the material constructions in another. But Seth says there are interconnections between universes, which he calls coordinate points. These points contain enormous energy potential and are essential in creating our universe even though they are extremely small — smaller than our elementary particles. The coordinate points are passive but can be activated by the emission of EE units. When a consciousness emits EE units, the mental activity expressed electromagnetically is projected into physical matter in an accelerated way. That is because the energy of the coordinate point slows down the EE units to our velocity. Thus, matter is an expression of thought or feeling.

The way the coordinate points work may be elucidated by the following metaphor. Picture a signal emitted from a radio station: it encounters an antenna that enables it to be received by a radio. The signal by itself is too weak to activate the radio's speaker, but the amplifier within the radio uses electrical power from a socket (or a battery) to increase the signal's energy to the level required to activate the speaker. In a similar way, the coordinate point acts as a power socket for the emitted EEs from the consciousness involved, producing matter by exploding EE units into physical events, literally breaking the space-time barrier.[4]

There are three types of coordinate points: *absolute* coordinate points (of which there are four) that interconnect *all* realities; *main* coordinate points, also sources of fantastic energy; and *subordinate* points, which are vast in number and affect our daily lives most directly. Seth made an interesting statement about these coordinate points:

> There is an ever-so-minute alteration of gravity forces in the neighborhood of all of these points, even of the subordinate ones, and all the so-called physical laws to some extent or another will be found to have a wavering effect in these neighborhoods. (Roberts, *Seth Speaks*, p. 76)

Since these points exhibit a slight alteration of gravity, they could be said to have a small mass in our universe. This raises the question of "dark matter," so called because it escapes the detection of telescopes. To be seen through the lens of a telescope, a body must emit radiation of some kind, and dark matter — estimated to be as much as 90% of the matter in the universe — does not do this. Dark matter is inferred because certain scientific observations seem inconsistent with the standard picture of the universe. For example, the only explanation for spiral galaxies spinning faster than predicted by theory is that they must be embedded in clouds of dark matter. What could be the origin of dark matter? Recall that physicists envision a great diminution of particles in the early universe

---

[4]This is related to Bohm's description of "active information":

> The basic idea of active information is that a *form*, having very little energy, enters into and directs a much greater energy. This notion of an original energy form acting to "inform," or put form into, a much larger energy has significant applications in many areas beyond quantum theory. (Bohm and Peat, p. 93)

Bohm's "form" is analogous to Seth's EE units, and his "much larger energy" is similar to Seth's coordinate points.

through collisions with antiparticles. Possibly particles that react more weakly with each other than electrons disappear more slowly and are left over to form dark matter. Other speculations have been offered, but the dark matter remains a deep and unsolved mystery. Equating dark matter with Seth's coordinate points is at least an intriguing idea.

Probable realities

More substantial similarities can be seen between another concept from Seth and Everett's "many-universes interpretation" of quantum theory (pp. 141–143). To circumvent the collapse of the wave function Everett postulated that every possibility splits off into a parallel universe, each with its own observer. That suggests that we have an improbable number of probable selves existing out there in other realities! Seth's comparable conception is that all mental acts emit EE units, but only some of them enter our reality. The others can be, and are, used elsewhere. In that way energy is not discarded or wasted. So, you may choose to stay at home and read a book rather than meet a friend for lunch, but once entertained as a thought, that luncheon date will take place in some other universe.

Our attachment to serial thought and our agreed-upon structure of three-dimensional reality make this exceedingly difficult to comprehend. Seth sums it up with the provocative statement, "Your slightest thought gives birth to worlds." Jeans also recognized thought as fundamental, suggesting that the universe is more like a great thought than a machine.

> . . . The universe can be best pictured, although still very imperfectly and inadequately, as consisting of pure thought, the thought of what, for want of a wider word, we must describe as a mathematical thinker. If the universe is a universe of thought, then its creation must have been an act of thought. Modern scientific theory compels us to think of the creator as working outside time and

> space which are part of his creation, just as an artist is outside his canvas. The old dualism of mind and matter seems likely to disappear, not through matter becoming in any way more shadowy or insubstantial than heretofore, or through mind becoming resolved into a function of the working of matter, but through substantial matter resolving itself into a creation and manifestation of mind. (Foster, p. 16)

Interaction between parallel worlds

Everett differs from Seth in feeling that the many universes are on parallel planes and therefore are mutually inaccessible — that you cannot be contacted or influenced by another probable self. That is not Seth's view. Before we explain the Sethian perspective, we shall examine a physics concept that will help us understand it more completely.

Feynman suggested an interpretation of quantum theory that has special relevance to the concept of parallel universes and their interaction. Feynman broke up the quantum waves into a sum of individual histories so that a particle moving from point A to point B travels as a wave taking *all possible* paths between the two points, with no one path taking priority over another. The wave function is then the simple sum of all the possible amplitudes of all the possible journeys. When Feynman added up all these paths, destructive interference due to their various phases caused the extremely nonclassical paths to cancel. The surviving paths were always within the vicinity of the classical path. As we have seen, here the distinction between time and space disappears. Called "sum-over-histories," Feynman's approach takes the realm of the wave function to be a sum of possibilities, and everything that *could* happen affects what actually *does* happen.

According to the Sethian view, each event of our three-dimensional reality is only part of the total event. Because we cannot grasp such multidimensionality, we select one portion and call it our reality, but all portions affect and influence all others. So, Seth is saying that these so-called parallel worlds do affect each other. This is not Everett's conception, but Feynman's sum-over-histories approach may make Seth's idea easier to understand.

The moment point and the total Self

Seth mentions often that he does not exist in any time framework, that minutes and hours do not have meaning for him as they do for us. However, since he is attempting to communicate with us, he tries to take our experience of time into account. Seth's conception of time is based on "the moment point," the present moment, in which all thoughts and all possibilities are explored in all their ramifications. That is, simultaneous actions and experiences are arranged in different associative patterns. For example, if Seth thinks of his Aunt Matilda, he immediately experiences her past, present, and future; her belief system; and all her strong emotions and motivations — all this in the twinkling of an eye. One can see why our senses would be overwhelmed with such a huge influx of information.

Seth says our brain is designed in a particular manner to enable us to perceive events arranged according to our perception of time. He sums it up as follows:

> Your brain gives you a handy and quite necessary reference system with which to conduct corporal life. It puts together for you in the "proper" sequences events that *could* be experienced in many other ways, using other kinds of organization. The brain, of course, and other portions of the body, tune into our planet and connect you with numberless time sequences — molecular, cellular, and so forth — so that they are synchronized with the world's events. (Roberts, *The Nature of the Psyche*, p. 180)

As Seth sees it, the concept of reincarnation as expressed in Eastern philosophy is essentially accurate although it is distorted because of our need to place events serially within the temporal dimension. Suppose we have had lives in various centuries. We assume that a tenth-century life is finished although presumably our behavior in that lifetime can have some effect on our present life. This is how karma is often interpreted. But Seth says that our present life influences the tenth-century life just as our future life influences this one, since all lives coexist in a timeless domain.

To imagine the relationship between the various lives, one other concept of Seth's must be briefly described, and that is the notion of the total Self. Seth sees each life as a projection from a multidimensional entity which he refers to variously as the total Self, the soul, or the entity. He describes it as follows:

> There is . . . a portion of you, the deeper identity . . . who decided that you would be a physical being in this place and in this time. *This* is the core of your identity, the psychic seed from which you sprang, the multidimensional personality of which you are part. (Roberts, *Seth Speaks*, p. 10)

From the perspective of the total Self, each person's energy is projected into three-dimensional reality and into numerous moment points simultaneously. While each projection seems to our three-dimensional ego to be isolated, a sense of wholeness is experienced by the total Self. Paradoxically, the total Self is both apart from us and yet we are in its three-dimensional form. All this is possible because of the concept of probabilities. That is, each life has many probable lives and our beliefs and actions in our present can alter our past lives, as well as the reverse. In this astonishingly complex process, the whole Self also develops and grows. Seth repeatedly emphasizes, however, that each individual is in total control of his or her own abilities and life challenges, because of the myriad probabilities available and our freedom to select among them. But the individual is also free to draw upon the energies and abilities of the whole Self. So, in some strange sense, all our selves are really simultaneous selves.

As we pointed out earlier, Schrödinger's equation remains the same if time is reversed. It is only when the wave function is collapsed that time reversibility is lost. This seems to indicate that the past is as open as the future. Can this state described by Schrödinger's equation be the same as Seth's moment point? That is, do we at each moment choose our past and our future? According to Wheeler's delayed choice experiment (p. 144–146), we do create the past as well as the future, since the past has no existence until it is recorded in the present. Such an idea clearly corresponds to Seth's notion that "past lives" are not fixed but can be and are constantly altered.

Creating our own realities in multiple dimensions

We have established that physicists view quantum systems as patterns in Hilbert space (p. 76). Seth's conception of creation uses metaphors similar in many ways to Hilbert space. However, Seth's views become more understandable if we examine mathematician Rudy Rucker's interpretation of Hilbert space as described in his book *The Fourth Dimension*. He starts with Wheeler's thought that "no elementary phenomenon is a phenomenon until it is an observed phenomenon," implying that the questions we ask profoundly affect the world we perceive. Thus, our four-dimensional space (adding time as the fourth) is only *one* way of organizing our sensations, useful mainly for describing the motion of matter. Taking perceptions — all we know for certain — as primary, Rucker considers other dimensions beyond position and time. In our sensory experience we also know the color of an object, whether it is hot or cold, how it tastes, etc. Rucker assigns a dimension to each of these aspects to create what he calls an infinite-dimensional "fact-space" in which each axis gives a particular sensory quality of an object. But since our perceptions are primary, each individual has his or her own fact-space.[5] Our total fact-space comprises everyone's perceptions, which are patterns in infinite dimensions. Rucker agrees that his fact-space is in many ways modeled on Hilbert space, but it is his enrichment of this concept that helps us understand Seth's vision more fully.

One of Seth's basic convictions, which he repeats over and over again, is that every gestalt of consciousness, regardless of size or complexity, creates its own reality. Seth discussed this in a session that included two other people besides Roberts. A glass was resting on a table nearby. Seth explained that none of them sees the glass that the others see; each person creates his or her own glass, from a personal perspective. Therefore there are physically three different glasses each existing in a different space continuum. "This can be proven mathematically," Seth says, "and scientists are already working with the problem, though they do not understand the principles behind it. . . . Physical objects cannot exist unless they exist in a definite perspective and space continuum." That continuum is created by each individual (Roberts, *The Seth Material*, p. 127).

To envision the idea that we all create our own reality, consider the following metaphor of a chess game. Two people in New York meet at one or the other's home to play chess. When one of the players moves to Boston, they decide to set up identical chess boards in the two cities and continue playing via e-mail. Their board configurations are identical, and the game progresses on separate chessboards. Similarly, Seth says we each have our own reality, but since we operate with the same neurological equipment, our realities are *largely* the same, and the game of life goes on. An artist and a

[5]Rucker also assumes that everything is in some sense alive and conscious, so presumably the sensations of all consciousness, not just human, are included. Since Hilbert space is infinitely dimensional, that is not a problem.

botanist will see a rose garden somewhat differently, but the fundamental sameness of their perceptions is such that they can discuss the garden in a meaningful way.

The concept of creating your own reality is similar to what Owen Barfield calls "configuration" in *Saving the Appearances — A Study in Idolatry*. Basically, Barfield does not see matter as causing sensations, but rather, as something constructed mentally from sensations, though we are totally unaware of this act of creativity. While Seth says that our creations are largely shared by others, Barfield refers to "collective representations," and the American sorcerer don Juan sees a "membership in an agreement."

Phase space, Hilbert space, Rucker's fact-space, and Seth's space all share the fundamental feature of multidimensionality. Whereas the metaphorical chess players communicate via e-mail, in actuality, all consciousness has available instantaneous communication through the level called the domain of the wave function, also referred to as level two, the implicate order, or Framework 2. Since this level exhibits a no-time attribute, each point in our universe is interpenetrated with all other points, and information transfer requires only focus, not time.

In the thought experiment involving the paradox of Schrödinger's cat[6] we imagine a cat in a closed box containing a radioactive atom and a device that can kill the cat when the atom decays. There is a fifty-percent chance of the atom decaying after one hour. What is the state of the cat at the end of the hour? According to classical physics, the cat is either dead or alive, but we cannot know which unless we open the box. In quantum theory, the cat is in a linear superposition of the two possible states. What does that mean for the cat? Looking into the box constitutes an observation

---

[6]Discussed more fully in *Bridging Science and Spirit*, pp. 243-254.

that collapses the wave function to produce either a live cat or a dead cat. Heisenberg would say the cat is in a state of potentia until we look.[7]

We have been assuming that a human observer is needed to bring one of the states into existence. But what if the cat itself is capable of this feat? And what about a single cell or even an electron? Where on the scale of life would this capability not exist?

Wigner and other physicists limit this ability to collapse the wave function to humans, but that brings up an additional problem called the "paradox of Wigner's friend." Suppose that Wigner is operating the experiment involving the cat, but instead of opening the box and making the observation himself, he asks a friend to do it. The friend opens the box, determines which state the cat is in, and reports to Wigner. Now, if Wigner collapsed the wave function, that function must include not only the cat but also his friend who made the observation. But we cannot reasonably conclude that only Wigner can bring this potential into reality — that only Wigner's consciousness is real and his friend is in a state of potential until he conveys the news to Wigner. Clearly, there would be some objection (particularly by the friend). We must assume, therefore, that the friend also is capable of collapsing the wave function. But suppose Wigner and his friend look at the cat together. Now who collapsed the wave function? A world of chaos would result if we were all going around collapsing each other's wave functions. It would be untenable to live in a world where your feelings create events for me. But if we accept Seth's and Rucker's concept of multidimensional space, then the cat, Wigner's friend, and Wigner each has its/his own space. They each collapse their own wave functions and thereby create their own separate realities, and everything else adjusts simultaneously. When

---

[7]Schrödinger's cat riddle dissolves the notion that the strangeness of the quantum world is confined to the micro level, showing that this paradox operates on the level of our daily lives.

the cat chooses being alive, in *its* reality Wigner and his friend alter their choices accordingly. On the other hand, if Wigner chooses a probability in which the cat is dead, in his reality, this is the case. Since everybody and everything has its own chessboard, these add up to a kind of Hilbert space which is, of course, infinite.

With this idea, we may have addressed one of Einstein's major concerns with quantum theory. Abraham Pais recalled a conversation in which Einstein asked Pais whether he really believed the moon has an existence only when we look. Had Seth been present, he might have said that Einstein's moon, Pais's moon, your moon, my moon, are all different moons. So if Einstein is not looking, only *his* moon is not there. If I am looking, *my* moon is there. It would have been interesting to hear Einstein's reply to such an explanation.

Suppose a group of very religious people jointly have a vision of the Virgin Mary, and a scientist in the crowd sees nothing unusual. Who is correct? Perhaps we can say that everyone is correct, but their chessboards are out of alignment. Furthermore, a person who persistently sees visions others do not see is usually considered outside social and psychological norms that define "sanity." The idea that we each create our own reality raises challenging issues in various disciplines.

Map and territory

In 1930 mathematician Kurt Gödel pointed out that we cannot start with basic assumptions and logically deduce a complete and consistent system because there will always be need for a truth that lies outside the system.[8] That is where Seth's Framework 2 and

---

[8]Gödel's incompleteness theorem states that there can never be a final system of mathematics since within a logical mathematical arrangement there are certain basic propositions or assumptions which cannot be proven true or false. If we widen the context to prove the basic propositions, new ones are created that are not provable within the new system. In short, we can never know reality in its entirety because it is infinite.

Bohm's implicate order come in. Seth states that we create our past and our future in the present moment. But the "we" is outside the physical universe and requires Framework 3 to activate Framework 2 to form Framework 1. In short, the "we" is in Framework 3. Attempts to include the total "we" involve more and more frameworks, finally approaching an infinite set or All That Is. Thus the concept of levels replaces Wheeler's need for bootstrapping.

Since Gödel's theorem applies to rational thought, a complete understanding of an infinite set cannot be acquired by deductive reasoning. In short, with our reasoning we make maps of the territory, but they should not be confused with the territory itself. Mystical experience may come closer to such reality than rationality, but even mystics are limited by Gödel's theorem in explaining the experience to the rest of us.

Seth says that even an electron is alive. How can we comprehend that notion? As we have noted several times, Newton codified the belief that matter is inanimate with his famous first law, or the law of inertia. This law states that matter cannot change its state without the action of an outside force; no inner volition is possible. Yet quantum theory says that indeed an electron does make choices among random processes, as we have seen from Einstein and Dyson, among others. We can now say that at the very least Newton's definition of "inanimate" is no longer valid if quantum theory accurately reflects reality. Since quantum theory is well-established, we can assume that in some sense all matter and therefore all energy is conscious.

In conclusion, it is obvious that Seth's conception of reality incorporates segments from a wide variety of interpretations of quantum theory. Most important is that both Seth and some physicists agree that our three-dimensional world is a projection from a higher-dimensional source. That source is at least partially described by the wave function.

NIGHT THOUGHTS OF A PHYSICS WATCHER

The evidence presented in this inquiry suggests that we are surrounded with, interpenetrated by, and immersed in the world of the wave function. But, as we have seen, the wave function is not in space-time as a physicist would define it but in what we are calling the hidden domain. One way of looking at it is that space-time is latent in this domain but is not actualized. Since we are completely immersed in the all-encompassing realm of the wave function, every point in our universe of possibilities is instantaneously connected with all other points; every point in space-time has access to all information in the world of the wave function. This is something like the popular saying, "Wherever you go, there you are."

During most of this book, I made an effort to ground my thoughts in interpretations offered by well-known and respected physicists. Now, with Seth as my inspiration, I shall give my ruminations a little more freedom. I will not stray too far from concepts in contemporary physics, but I may go a bit further than some readers will care to follow. I invite you to ignore whatever offends you and to reflect upon those ideas that are appealing.

16 Light thoughts

One of the most striking features of both classical and contemporary physics is the role played by light. It was, in fact, the behavior of light that stimulated the ideas that gave birth to quantum physics. A brief review of scientific theory related to light in the past century will show how this is so.

Faraday and Maxwell conceived of light in a new form called fields, which carried energy but were disembodied. Oscillating electric and magnetic fields were seen to determine the motion of light through space. Maxwell's equations indicated that these electromagnetic waves constitute a spectrum of frequencies of which visible light is only one small portion and that the velocity of light is fixed and is not dependent on the motion of the observer. This idea led to Einstein's special theory of relativity and a completely new understanding of matter — that matter and energy are interchangeable. With quantum theory and the two-slit experiment (pp. 51–55), physicists were forced to the reluctant conclusion that light is both particles and waves — an apparent contradiction, which has, however, been experimentally confirmed.

In quantum theory, a particle is considered a quantized state of the field. So a field is spread all over space and yet, in some peculiar way, it exhibits a quantum of energy, the amount of energy being proportional to the frequency of the field. In gravitational or electromagnetic fields an infinite number of waves is possible, adding up to an infinite amount of energy.

The ocean of light

Bohm equates this energy pool with the domain of the wave function and describes it as a vast ocean of light. In this ocean of light, there are no separate points. Light does not travel from one place to another, it just *is* — a continuous, unbroken whole. From relativity theory we know that only massless particles can travel at the speed of light, so if the hidden domain is an ocean of light, it cannot exhibit the property of mass. How, then, does the mass of our world relate to this massless realm? We know that an electron and a positron contain a certain mass and therefore energy. But if they interact and destroy each other in the process, they are survived by a photon of light with the required frequency. So one can say that different rays of light can produce particles (and therefore matter) when they interact properly. Thus, the hidden level can be seen as an ocean of light which, under the right conditions, can form into particles.

Where is the hidden level? Certainly it cannot be found within space-time parameters, so it does not have a location in our usual sense of the term. Perhaps there are limits in human neurology that preclude our perceiving this domain. According to the Biblical story of Job, God says, "Have you comprehended the vast expanse of the world? Come, tell me all this, if you know. Which is the way to the home of light?" (Job 38:18,19). Apparently Job did not comprehend the ocean of light any better than we do.

Other universes

Plato compared our understanding of the world with that of people living in a cave with their backs to the opening. Outside the cave is the true reality, represented by light, but the people inside see only the shadows on the cave wall. They develop elaborate systems of science and philosophy that analyze the shadows in great detail, while the true reality lies beyond their sensory grasp. In the same way, we are fascinated with the three-dimensional world around us, unaware of its source — a much greater reality.

Similarly, Seth says that there are many gradations of matter that we do not perceive because they are constructed of particles that move faster than the speed of our light. That is, there are other universes differentiated from each other by the velocity of light. Matter composed of other velocities of light would not be perceived since we are neurologically designed to function specifically in this universe. So, while other universes may, so to speak, share our space, they would completely escape our notice.

Accessing the hidden domain

Because our physical senses and instruments cannot detect phenomena outside our three-dimensional universe, we are limited in our access to other levels. This does not mean, however, that we do not experience the hidden domain or that access to it is impossible. We reach it not with instruments but by going outside our usual sensory pathways. In altered states of consciousness other portions of the velocity-of-light spectrum seem to be available to us. Methods of achieving altered states are offered through such disciplines as yoga, meditation, prayer, as well as through psychedelic drugs and by various other techniques.

An altered state of consciousness common to us all is dreaming, in which information is conveyed by symbols and images. According to Seth, the everyday objects around us are also symbols created by our minds, but symbols of our waking life. In dreams, the individual can display a wider range of information than is usually available since the objects of the dream world are gestalts of EE units outside our velocity of light. The dream world, in short, is another universe with altered basic constants. To study this universe, we do not need particle accelerators and bubble chambers, but a proficiency in lucid dreaming might be helpful.[1] A lucid dreamer is aware that he or she is dreaming, often can control activities within the

[1]For more detailed information, see *Lucid Dreaming* by Stephen LaBerge, a scientist at the Sleep Research Center at Stanford University.

dream, and thus may gain knowledge of this other realm that is ordinarily unobtainable.

While Seth agrees with a number of dream researchers that we tend to translate experience into so-called dream objects, he does not see that as our only method of dreaming. During a portion of our dream time, Seth says we communicate with deeper parts of our being, which we might term the soul or higher self. This contact, however, is not translated physically at any velocity, and therefore the experience is not remembered when we wake. That is not to suggest that this nonphysical experience is wasted; on the contrary, the information gained may help form dream objects for subsequent dream states, even that same night. Seth calls this a depth experience, or the predream state. The following excerpt describes this state in probabilistic terms:

> In the predream state you directly encounter a reality in which . . . probabilities exist all at once to your perception. In a dazzling display you are aware of such events from infinite perspectives. Consciously you could not grasp such information, much less act upon it, nor could you maintain your particular, unique, psychological stance. You still take advantage of that level of being, however, using that immeasurable data as a basis to form the reality that you know. (Roberts, *The Nature of the Psyche*, p. 153)

Light in religious scripture

Light has always been important in accounts of mystical experience and other religious writing. In fact, many sources have noted that light is what unites us and makes us one with God. The ancient Egyptians saw light as the gaze of the Sun-God Ra — light was thought of as God seeing. According to the *Tibetan Book of the Dead*, the dying person first encounters what is described as the primary clear light, taken to be the ultimate reality, or God, as in the following excerpt:

> Thine own consciousness, shining, void, and inseparable from the Great Body of Radiance, hath no birth, nor death, and is the Immutable Light, Buddha Amitabha. (Hooper and Teresi, p. 321)

Light is also the primal element in the Judeo-Christian tradition. "God said, 'Let there be light'" (Genesis 1: 2-3), and out of the primeval chaos came illumination. Gradually, this light that was called forth became identified with God, as in, "The people who walked in darkness have seen a great light" (Isaiah 9:2). "God is Light" passages in the Bible are well known; for example, in the First Letter to John we have, "God is light, and in Him there is no darkness at all" (1 John 1:5). Also in the apocryphal and rabbinic literature, light takes a prominent position as an "everlasting" entity. The gnostic Christian tradition sees light as the basis of reality, illustrated in the following quotation from the Gospel of Thomas:[2]

> Jesus said: I am the Light that is above them all, I am the All, the All came forth from Me and the All attained to Me. Cleave a piece of wood, I am there; lift up the stone and you will find Me there. (Wilber, p. 62)

God and light are associated with love and knowledge, whereas the devil and darkness are associated with evil and ignorance. "God saw that light was good, and He separated light from darkness" (Genesis 1:4). Light and darkness have represented good and evil in much religious symbolism. Also, most traditional myths assume that originally there was darkness and then God brought forth light. Zoroastrianism uses this dualism to picture a struggle between the forces of light and darkness, in which the forces of light are the victors. In Christianity,

---

[2]Found in the 1940s at Nag-Hamadi in Upper Egypt, this gospel is generally believed by scholars to have circulated in Egypt but to have originated with a Jewish-Christian community closely tied to James, the brother of Jesus.

this dualism is represented by heaven and hell.

Seth, on the other hand, does not accept this dualistic conception; he completely rejects the notion of opposites. This point of view is clearly stated in the following:

> Opposites have validity only in your own system of reality. They are a part of your root assumptions, and so you must deal with them as such.
>
> They represent, however, deep unities that you do not understand. Your conception of good and evil results in large part from the kind of consciousness you have presently adopted. You do not perceive wholes, but portions. . . . The effect of opposites results, then, from a lack of perception. (Roberts, *Seth Speaks*, p. 406)

So, if Seth is correct, when we see what we call darkness, the absence of light is due to our lack of perception. That is, there is light within darkness but we are unaware of it. Assuming this to be true, if we probe the darkness, we should discover light. That is, light does not have an opposite; darkness is a lack of perception.

Seth's conception resonates with Bohm's view that space is not a vacuum but a plenum. The plenum is an ocean of light that is not measurable by our instruments nor detectable by our senses. All this is analogous to Maxwell's discovery that light is really a frequency spectrum, though our vision is sensitive to only a small part of the spectrum. If our senses detected only x-rays, then what we experience as darkness would be something very different from the darkness we know. Similarly, since our instruments do not detect faster-than-light radiation, we define light in a particular and limited way. In the broad sense, there is no darkness, just light. The Gospel of John says it this way: "The light shines on in the dark, and the darkness has never mastered it" (John 1:5). The Chinese philosopher of the sixth century B.C., Lao-tzu, identifies darkness with the source, what we call the hidden domain. These are his words:

> Mystery and manifestations arise from the same source. This source is called darkness. . . . Darkness within darkness, the gateway to all understanding. (Zajonc, p. 325)

Light and the near-death experience

Reports from people who have been on the brink of death, or were thought to be clinically dead and then revived, include unusual events connected with a world beyond our everyday world, many involving encounters with beings of light. The usual scenario is something like this: At the moment of apparent death, the person undergoes an out-of-body experience resulting in a complete change of perspective so that his or her own body is viewed from above. Then a portal opens, and the person is propelled through a dark tunnel. At the end of the tunnel is the being of light. (Some people also describe seeing beautiful cities of light as well as meetings with luminescent departed friends.) However, this is not ordinary light; it is described as much brighter than earthly light, yet is not blinding. Many who have experienced it say they feel drenched with love and peace, and they use the words "light" and "love" interchangeably. Interestingly, Bohm said that our world of matter is like waves on the ocean of light, and that contact with the depth of this ocean produces a profound peace and serenity compared to the more turbulent surface (Weber, p. 49).

All That Is

Contrary to our common perception, light is a continuous and undivided whole; it does not exist in time nor does it travel through space. Rather, light is the basis of all levels of reality including the hidden domain we have been examining. Perhaps surprisingly, this startling conception coincides with the utterances of mystics of all ages. Seth uses the term "All That Is" for what we call God, so there is no basic conflict between the Biblical "God is Light" and our conclusion that All That Is is light.

17 Timeless thoughts

Two domains, one reality

The sense of the passing of time is a common feature of human experience. We not only feel this passage personally, we also observe it in our surroundings. In spite of the fact that we all know intuitively what time is in our lives, some philosophers and scientists question its very existence — not the human experience of time but its existence independent of the observer. That is, is time an objective phenomenon? Does it exist in its own right? Perhaps it is as Einstein remarked, that time is one of the ways human beings order experience.

Speculation about time goes back to such Greek philosophers as Parmenides, Zeno, and Plato. Plato's view, in the context of the cave metaphor is particularly interesting: the shadow world is in a time domain, while the universe of light outside is timeless. Plato uses the term "Being" to designate the primary world of light and "Becoming" for the secondary shadow world. The shadow world comes into existence and then ceases to be; the universe of light just is.

All psychic experiences must be described (though not experienced) in terms of matter and motion. Since we seem, by our language, to be restricted to these concepts, we see growth in terms of material change, with time a necessary component. The ocean of light within which all exists is itself timeless, but the waves on its surface — unfolding into our reality — exist in space-time. In the same way, Seth describes growth in our physical universe as taking up more space and projecting into time, whereas in the inner universe growth exists in terms of what he calls value fulfillment (the analog of growth in the hidden domain), and does not imply any temporal or spatial extension.

The idea of having domains both in and out of time is reflected in the formalisms of quantum theory. In our daily lives we know time to be irreversible; the past seems fixed and the future seems open. In physics, however, the direction of time seems to be in most cases immaterial. That is, all the major equations of physics — Newton's, Maxwell's, Einstein's — are symmetrical in time, working just as well, completely unchanged, whether we use a minus t (time) or a plus t.

When we turn to quantum theory, however, a peculiar situation arises in reference to time. As we have seen, the Schrödinger equation gives us a completely deterministic time evolution of a quantum state; that is, a quantum state is described by a wave function, and Schrödinger's equation tells us how it evolves with time. If the Schrödinger equation were the complete description of reality, all the indeterminism that troubled Einstein would be eliminated since at this point quantum theory is similar to Newton's, Maxwell's, and Einstein's equations in that the arrow of time is reversible. But the reality we know and love is created by what we call the *collapse* of the wave function — and it is precisely this that introduces the arrow of time and makes our universe come alive. As we have seen, the connection between the time-reversible mode of Schrödinger's equation and the creation of a particular event is an act of observation. So quantum theory is both deterministic and without an arrow of time and *also* nondeterministic with an arrow of time. Before the collapse, the wave function exists in a domain outside space-time; after the collapse, time is a factor. So space-time (and therefore time) seems to be created by an observation by something outside space-time.

In terms of space, let us return to the idea of reality being like waves on an ocean. The electron when observed is like a particle in the ocean with a specific location; in between observations, the electron is spread throughout the ocean, with the potential to be anywhere. Again, quantum theory describes reality as made up of two domains, one timeless, spaceless, and completely continuous or interpenetrated, the other timed, with matter occupying different areas of space. We could equate Plato's world of light with the domain of the wave function before collapse and the world of shadows with our three-dimensional experience of matter, motion, space and time after the collapse.

The thought that our everyday macroscopic world is embedded in a timeless world of potential is anathema to many scientists and philosophers, especially if our world is seen as a projection from a more primary domain and the projection is created by consciousness. Therefore, some physicists survey the problems of quantum theory (especially the measurement problem) and declare it incomplete. After all, they say, it is just a theory — our particular map to describe certain experimental data. Their point of view, it seems to me, contains a major flaw, which has to do with a lack of appreciation for the significance of Bell's theorem, which requires all of reality to be linked up superluminally — making the universe an undivided whole.

You might counter that, after all, Bell's theorem is also just a theory. But Aspect's experiments confirmed it, and therefore we conclude that reality must be nonlocal (p. 92). But what about the theory of relativity, which explicitly states that there can be no energetic signals faster than light — and also has been confirmed by experiment? The inescapable suggestion is that reality has both local *and* nonlocal aspects; that there exists a more fundamental level than our macroscopic world, which corresponds to Heisenberg's potentia, Bohm's implicate order, Seth's Framework 2, and the world of light outside Plato's cave. All of these denote the timeless level of reality, the world of the wave function before collapse — the level from which, through the act of observation, our three-dimensional world is created.

As we have seen in Chapter 15, we create time and space by selecting potential structures and events from Framework 2 and unfolding them in Framework 1. Framework 1, our everyday world of matter in motion, has an arrow of time; relativity and entropy are operable, all motion and energetic messages are limited to the velocity of light and below, and entropy increases. Framework 2 can be identified with Heisenberg's potentia, the world of the wave function before collapse, the superposition of all possibilities that can be unfolded and exhibited in Framework 1. In Framework 3 the choice is made as to exactly which potential will be elevated to actuality. Some physicists say the selections are made at random — and thus believe that God really does play dice with the world. These scientists feel that we know nothing about such a third level and, even if it exists, it certainly does not follow any of our known laws. Einstein referred to it as the level of hidden variables. On the other hand, physicists such as Bohm, Wigner, and E. H. Walker call this the level of consciousness, which draws from the domain of possibilities to create the level of events. If we look at these levels as different aspects of one reality, then Bell's theorem and Einstein's relativity can live together in harmony.

Psychological time

Seth introduces another aspect of time that has relevance to contemporary physics in his conception of inner senses. Seth's suggestion is that everything in our world (Framework 1) is the materialization of something that exists on another plane. Just as our regular senses perceive and create the outer world, so do our inner senses perceive and create an inner world. Since we physically exist in Framework 1, it is natural for us to place our awareness in that plane, but as we grow psychically we can learn to use our inner senses to a greater extent and thereby increase our scope of activity in other planes. It is all a matter of focus.

One of these inner paths Seth calls Psychological Time, or, as Jane Roberts called it, "Psy-Time." Psy-Time is a natural pathway that gives us easy access between inner and outer worlds. Earlier peoples used this inner sense with relative ease, but we have lost some of that facility since we are so heavily focused and invested in outer reality. According to Seth, however, we can use this pathway even while we are consciously awake, but we must become aware of its presence and be trained in its use.

To understand the relevance of Psy-Time to physics, we need a brief review. Recognizing that Einstein's space and time were really a single space-time domain, Minkowski introduced time as a fourth dimension described by imaginary numbers. Imaginary time performs like ordinary time but has no direction, and most physicists consider it a mathematical expression without correspondence in the real world. If one defines the real world as Framework 1 only, then this position has merit. On the other hand, if one concludes, as we have, that Framework 1 is just one aspect of reality, then perhaps imaginary time is the more fundamental concept. Our everyday subjective time is then only a construct we use to order events.

Recall that Minkowski multiplied the time axis by the speed of light to spatialize it and thereby create space-time. The movement of time from present to future is visualized by many physicists as the entire universe moving down the imaginary time axis (*not* through spatial dimensions) at the speed of light. As we have noted, the velocity of light defines a universe in a very real way, by determining the natural units of length, time, and mass, which define our universe (p. 103). Other universes, with different velocities of light and therefore different natural units, are also going down the imaginary time axis, which we can picture as a spectrum of different universes linked only in Psy-Time. If the imaginary time axis in some fashion contains all other realities, then each can be defined by a point on the axis determined by the velocity of light it uses.

Seth says:

> Psychological Time is a natural pathway that was meant to give an easy route of access from the inner world to the outer, and back again, though you do not use it as such. Psychological Time originally enabled man to live in the inner and outer worlds with relative ease. . . . As you develop in your use of it, you will be able to rest within its framework while you are consciously awake. It adds duration to your normal time. From its framework you will see that physical time is as dreamlike as you once thought inner time was. You will discover your whole selves, peeping inward and outward simultaneously, and know that all divisions are illusion. (Roberts, *The Seth Material*, p. 280)

Psy-Time gives us access to alternate realities and to our dream world. We can change our sense of time and even attain states in which time is completely absent. In that way, we can know that there is some portion of reality where time does not exist. If, in such a reality, consciousness is not required to materialize its thoughts, then no energy is required, and perhaps that is where tachyons attain an infinite velocity by losing all their energy. This state of zero energy and infinite velocity, where place and time have no meaning, is referred to as the transcendent state of the tachyon.

To conclude, space-time is a concept we use in our three-dimensional reality. There may be other realities with other states of time, defined by different fundamental constants. Finally, there may well be other realities, like Seth's, that do not use time at all.

18 Creative thoughts

Seth describes how our world is created as follows:

> Some feelings and thoughts are translated into structures that you call objects; these exist, in *your* terms, in a medium you call space. Others are translated instead into psychological structures called events, that seem to exist in a medium you call time.
>
> . . . Your inner experience is translated in those terms. (Roberts, *The Nature of Personal Reality*, p. 12)

It is fairly well accepted that mind can influence matter. We know that our moods and emotions have an impact on our bodies in numerous ways. But as a society we do not accept the premise that mind *creates* matter — an idea basic to Seth's teachings and a view that resonates with quantum theory. According to this perspective, not only are sentient beings capable of this remarkable creativity, but other gestalts of consciousness have this same ability to a greater or lesser degree. We humans are made of the same units as a rock or a star, and in a gigantic cooperative effort, through our innate unconscious ability to transform electromagnetic energy units into physical objects, all consciousness joins together to make the forms we perceive. Seth says that by transforming our thoughts and emotions into physical form, we learn how to use our mental energy. This mental energy is light, but light of a higher velocity than that in our three-dimensional world. When we perceive the physical environment we have created, we get a clearer picture of our inner development since the outer world is not independent of us but the materialization of our own mentality.

The Sethian explanation of matter formation, like Cramer's transactional interpretation, includes waves (EE units) meeting in time, with the brain interpreting the information in the forward (retarded) wave while the mind creates the object with the backward (advanced) wave. This reflects the mental and physical qualities of our being, associated with the mind and the brain. Neuroscientist Candace Pert employs a wonderful metaphor for the relationship between the mind and brain. Noting that scientists soon will be able to create a color-coded wiring diagram of the brain, and asked if such a diagram would account for consciousness, she replied that it would not.

> Just as a person may totally understand a television set — can take it apart and put it back together again — but understand nothing about electromagnetic radiation, we may be able to study the brain as input-output: sensory input, behavior output. We make maps, but we should never confuse the map with the territory. I've stopped seeing the brain as the end of the line. It's a receiver, an amplifier, a little, wet minireceiver for collective reality. (Weintraub, p. 121)

Seth seems to be in agreement with Pert. He says:

> The brain organizes activity and translates events, but it does not initiate them. Events have an electromagnetic reality that is then projected onto the brain for physical activation. Your instruments only pick up certain levels of the brain's activity. They do not perceive the mind's activity at all, except as it is imprinted onto the brain. (Roberts, *The Nature of the Psyche*, p. 180)

Here is how this process of creating reality may work. The mind projects thoughts outward toward the bank of physical potentials. These projections, in the form of EE units, travel faster than our light. Because they are going faster than light, according to our time sense they are waves going backward in time, like Cramer's wave function traveling into the past. This wave "selects" among the potentials and gives an objective existence to the "out there." The objective event then sends a wave forward in time to our brain, the physical aspect, which then perceives the event. The advanced wave effects are cancelled by being out of phase, thus preventing us from being aware of how we accomplish this remarkable feat of creativity. To use Cramer's word, we use the future to create a "handshake" across space-time and thus bring into reality the event we desire.[1] When looked at in this way, all events are initially inner events and it is through this backward and forward process that the home of the wave function is affected and space-time events are created. Astonishing as it may seem, we are even more creative than this, for each selection multiplies the actions possible for all others.

Let us return to Comfort's analogy of a computer game in which the black box of electronic circuitry represents the world of the wave function, the screen represents our actualized three-dimensional world, and the player represents our consciousness. Seth says that our material environment is a projection that allows us to examine our innermost thoughts, feelings, and beliefs. The player operates like a remote control, using electromagnetic radiation to convert his intentions into signals. These signals — Seth's EE units — employ not our familiar photons but tachyons, which go faster than light. They have access to the power socket of the computer — analogous to Seth's coordinate points — from which they gain the energy necessary to select the intended probability. The screen of each person's computer is separate and individual, but through a superluminal link (described by Bell's theorem), we all keep our screens alike. Those who do not (or cannot) follow this rule are often either ostracized or institutionalized.

The computer screen — our material world — is necessary to help the players hone their skills in an efficient way. That is, we

---

[1]Cramer would not agree that an observer is necessary in quantum theory. We have borrowed some of his ideas to illustrate Seth's views.

express ourselves through matter, and material life provides consciousness with an excellent method of evolving, growing, and attaining value fulfillment. Just as we may record our thoughts on paper (as I have) by choosing appropriate words and arranging them to communicate with others, in a similar sense, in our creation of physical reality, matter is like materialized thought.

There are many ways for consciousness to express itself. On the primitive level of the amoeba, in all likelihood, objects are not created; at the other end of the spectrum, according to Seth, there are gestalts of consciousness that do not share our human need for manifesting our thoughts in a physical dimension. Even within a framework of materializing thought, this manifestation of matter can be accomplished in innumerable ways. By utilizing different values for the velocity of light, an infinite number of material subtleties are available, so that within our own space an enormous range of universes can be employing different levels of matter. An equal number of time orders is also possible. Since the consciousness of the creator in each universe is tuned to a particular velocity of light, all these universes can coexist and operate without encroaching upon each other. It is as if each universe is tuned into one point on an infinite radio dial (on which velocity is substituted for frequency) with all other stations effectively blocked out. According to Seth, "bleed-throughs" that occasionally occur may account for our experience of such phenomena as UFOs.

In the paradigm presented here, the wave-particle duality takes on a broader meaning than in quantum theory. Recall from the two-slit experiment that when we are looking (or measuring), the electron becomes a particle. When we are not looking, it appears as a wave. Instead of using the term "looking," suppose we use the term "creating." We create the electron not in a literal, mechanical way, but we request that the electron particularize itself so that it can be perceived in our world. This being a very cooperative reality in which we exist, the electron complies. In an analogous manner, the entire human race uses us as individuals to carry out its extensive (and often elusive) intents and purposes. But paradoxically, we cooperate in the process in order to pursue our own personal goals.

The point being made here is that we humans also have a wave and particle aspect; our consciousness is represented by a waveform, and our body by a conglomeration of particles. Our waveform (consciousness) mixes and merges with those of other people and entities and in so doing our identities grow and expand. Thus, the universe is permeated with creation — from the electron coming into being to our own purposes, which ultimately further the creative evolution of All That Is.

19 Conscious thoughts

The mystery and the theories

Consciousness is a mystery. In spite of our great scientific advances, we have no theory of consciousness. I am not saying we do not have a "good" theory, but that we have *no* theory, good or bad. Science has tackled many other mysterious phenomena such as electromagnetism, for example, and eventually measured them. But consciousness has not proved accessible in this way. It is rather similar to the wave function in that it does not carry energy as we define it. Yet what is more familiar to us than our consciousness? We all experience perception, cogitation, and rumination. When asked how we do these things, we might reply that we are using our consciousness, which seems to be the essential feature of our minds. No one doubts that each human being possesses consciousness, but how about a tiger, a tree, an amoeba, or an electron? Without a theory or context, such questions are difficult to answer scientifically. However, with Seth's help, I shall be bold enough to address the subject, even though we can only partially understand this complex phenomenon.

The observer was introduced into physics with special relativity theory, when Einstein showed that something as ordinary as simultaneous events depends on an observer's frame of reference. In quantum theory, the observer, with a measuring instrument, is essential for the creation of events. Schrödinger's equation describes a quantum system with all its possibilities, which evolve with time but are incapable of manifesting in our world; the solution to Schrödinger's equation, the wave function, is passive. Von Neumann concluded that the midwife, or that which delivers reality, is the mind of the observer. Mind immediately brings up consciousness, because we assume only a conscious observer has a mind. Yet one can search through all the equations of physics and nowhere find a *c* for consciousness or *m* for mind. The equations of quantum theory successfully and beautifully relate the pointer readings and computer readouts of instruments, but they do not tell us how to get from Schrödinger's equation to our real world except to say that it is done by an act of observation. According to physics texts, we seem to spring from a world of potentiality to one of actuality in one miraculous leap. Though we may accept that consciousness is in a sense the choreographer of our world, we have not a clue about how the dances are selected.

The idea that electrons have a form of consciousness is found not only in quantum theory but in various philosophical systems. Theosophy views human and all other beings as emanations from a Universal Ground, and similar ideas are found in Hinduism, Buddhism, and even Christianity. Ken Wilber discusses this in his book *The Spectrum of Consciousness* (pp. 61-63).

> [The] "Christ-only" experience is formally indistinguishable from that of "Mind-only" of the Buddhists or physicists, and moving on to Hinduism, these are both formally indistinguishable from the core "doctrine" of Vedanta that Reality is Brahman-only. (Wilber, p. 62)

The philosopher William James described different types of consciousness in his writings. He recognized that our normal waking consciousness, consisting of separate moments of awareness exhibiting continuity through memory, with the brain and sense organs maintaining focus, is but one segment of all the potential forms of consciousness. Dreaming and hypnotic states are other examples, as are mystical experiences and trances.

The consciousness of waking life is generally focused on a

narrow range of experience, while other equally valid dimensions tend to be ignored. By restricting our experience to exclusive identification with the ego (see discussion on p. 171), we deprive ourselves of access to other portions of our consciousness (sometimes referred to as the unconscious) which are enriching and necessary for our existence.

Jung considered instinct to be contained within the unconscious, defining the unconscious as the totality of psychic phenomena of which the individual is unaware (though we all know that unconscious processes can alter the behavior of an individual significantly). He used the concept of a threshold of energy as that which separates the conscious from the unconscious. Jung also differentiated between a personal unconscious, which includes such things as lost memories and repressed thoughts and feelings, and the collective unconscious, where he placed his famous archetypes. Having made these distinctions, Jung asserts that he does not see the unconscious and conscious as truly separated. The following statement makes that clear:

> We must . . . accustom ourselves to the thought that conscious and unconscious have no clear demarcations, the one beginning where the other leaves off. *It is rather the case that the psyche is a conscious-unconscious whole.* (Enz and von Meyenn, p. 153)

Focus

One of Seth's most interesting statements is that he is aware that he is not his consciousness, but that consciousness is an attribute he uses. We have been assuming that consciousness is present in all energy gestalts,[1] that it is pervasive throughout reality — from

---

[1]Seth attempts to define who we are in terms of spirit, but he does so with trepidation, knowing that the meaning of the word is often distorted or misunderstood. So for the sake of discussion, let us use the term energy gestalt, remembering that it implies spirit, or an independence from physical form.

electrons to humans to All That Is. But each energy gestalt has a different kind of consciousness, and what distinguishes them is focus.

Seth says within the energy gestalt of an individual, consciousness is like a spotlight which can be turned in many directions. In sleep, for example, the spotlight is turned away from our physical reality and instead illuminates our dream world. Our normal waking consciousness is not extinguished when the focus is changed, but it becomes dimmer than usual. Apparently a similar situation occurs when one is close to death. Seth says a dying person who over-identifies with the body as the expression of its waking consciousness can become panic-stricken with fear that death of the body will extinguish the entire consciousness of the energy gestalt. *But that is not the case*. The consciousness spotlight may dim as death approaches but is preparing to shine in another dimension of reality.

Another Sethian concept is that of alternative focus, in which the spotlight of consciousness is turned slightly off of its habitual direction, like looking out the corner of one's eye rather than straight ahead. This allows one to relax the ordinary awareness of time and space. Some obvious means of altering focus are meditation, out-of-body experiences, and dreaming. But slight alterations of focus happen unnoticed quite often, as for example, in reverie and daydreaming. These are creative moments when normal consciousness slips or lets down its guard somewhat, allowing energy in from other areas. In these so-called lapses, one can, if properly tuned and trained, perceive other realities. According to Seth this can happen as often as fifty times an hour. However, in most cases the conscious mind does not accept and record this information in its memory system.

In *The Emperor's New Mind* Penrose discusses such flashes of insight, using the example of an incident related by mathematician Henri Poincaré. In his search for what are known as the Fuchsian functions, Poincaré had reached an impasse after much conscious thought. Then while he was boarding a bus, an inspiration entered his mind. He described it as follows:

> At the moment when I put my foot on the step, the idea came to me, without anything in my former thoughts seeming to have paved the way for it, that the transformations I had used to define the Fuchsian functions were identical with those of non-Euclidean geometry. I did not verify the idea; I should not have had time, as upon taking my seat in the omnibus, I went on with a conversation already commenced, but I felt a perfect certainty. On my return to Caen . . . I verified the result at my leisure. (Penrose, p. 419)

Penrose comments:

> What is striking about this example . . . is that this complicated and profound idea apparently came to Poincaré in a flash, while his conscious thoughts seemed to be quite elsewhere, and that they were accompanied by this feeling of certainty that they were correct — as, indeed, later calculation proved them to be. (Penrose, p. 419)

That is, Poincaré did accept this moment of alternative focus, and so it became part of his memory system.

The backdoor metaphor

Jane Roberts compares the conscious mind to a house of awareness. The front door is the main entrance, used by most people who come to the house, and it is where mail and other messages are most often received. More care is taken with the lawn and bushes in the front of the house, and the best curtains hang in the front windows. The front door is our face to the world. But the house of awareness also has a backdoor, an entrance to other worlds. All of us have sensed at one time or another messages or intuitions left at the backdoor of the mind, but largely we are taught to disregard this door, to deny entry to what knocks there. So, for most of us, unless we make a committed effort to turn to that door, and allow it to open, we limit ourselves to the more acceptable front entrance.

We know how to read the kind of information — much of it verbal communication — that comes through the front door. But the messages of inspiration, intuition, and insight left at our backdoor are often expressed in symbols and images we do not understand rationally. Furthermore, the information coming in the backdoor may contradict the messages of the front door, those which are sanctioned by authority and social convention. In short, in our materialistic world, the portion of the mind that is suited for the translation of backdoor messages is untrained, unused, and often not socially accepted. Seth describes the situation as follows:

> Certain portions of the brain *seem* dominant only because of those neural habits that are adopted in any given civilization or time. But other cultures in your past have experienced reality quite differently as a result of encouraging different neural patterns, and putting experience together through other focuses. (Roberts, *The Nature of the Psyche*, p. 181)

The prejudice regarding the backdoor may be diminishing somewhat. Perhaps Feynman had a version of the backdoor in mind when he said, "If we want to solve a problem that we have never solved before, we must leave the door to the unknown ajar" (Jones, p. 34).

One way to envision the enormous range of consciousness is to imagine a radio dial with an infinite spectrum. Usually we are tuned to our favorite frequency — physical reality — but we sometimes hear bleed-throughs from other stations. Some of us can tune to a wider range of frequencies, and some of us have the ability to listen to more than one station at a time. Perhaps with special training or with more openness, we could learn to broaden our individual spectrum. In principle, we are all capable of receiving all stations.

Identity

According to Seth, consciousness units come together in meaningful patterns to form inviolate personal identities. With this subject it may appear that we have left the boundaries of physics far behind, but Seth's description shows us that that is not necessarily true. Cautioning us that these matters are difficult to understand, Seth explains in the following passage how consciousness units form patterns of identities.

> [CUs] move faster than the speed of light. They can be in more than one place at one time. They can operate in a freewheeling fashion, as identities in themselves, or as "psychological particles."
>
> They can also operate in a wavelike fashion, flowing through other such particles. They can form together into endless, infinite combinations, forming psychological gestalts. Certain portions of these gestalts can then operate as psychological particles in time and space, while other portions operate in a wavelike manner outside of time or space.
>
> These represent the unconscious elements of the psyche, which become "particleized" in physical existence. (Roberts, *The Afterdeath Journal of an American Philosopher*, p. 9)

Relating this statement to concepts we examined regarding the wave function, and assuming that Seth's psychological particles are the same as electrons, we can see that the pattern of identity of an electron can operate in a wavelike manner outside of space-time or as a particle in space-time. We know that the wave function, which describes the quantum system represented by the electron, is also outside of space-time. It does not use three-dimensional space in that it can be in more than one place at a time; in other words, it is located in a multidimensional space. It evolves with time — not the time we experience, which moves only from past to future and results from the collapsing wave function — but with imaginary or reversible time. Since we are equating energy and consciousness, whether or not we can measure the energy with our three-dimensional instruments, then both the particle and its wave function are conscious, with the wave function corresponding to the part of the psyche we call the unconscious. That is, our bodies can be considered explications of the wavelike portions of our identity, and the so-called unconscious portion of us becomes material by the processes delineated throughout this book.

As you recall, the concept of phase entanglement tells us that when one quantum system represented by a wave meets a second quantum system represented by a wave, part of the wave function of the first system goes off with the second system wave, and vice versa. More accurately, their phases become entangled, and this interaction provides, henceforth, an instantaneous connection between the two.

Similarly, wave formations flow together in currents, mixing and merging, so that components constantly change, yet identity remains inviolate: each wave, since the wave function is linear, can interact and depart intact. Due to phase entanglement, your identity remains even though you as an individual change and grow. In this awesome process, there are no boundaries or limitations to the self except the ones an individual chooses. We are at once separate individuals and yet totally unified with all other consciousness; we must, therefore, respect both our individuality and our unity. As Seth said to Jane Roberts:

> Bits of your consciousness . . . go out through these books. I am not speaking symbolically. These portions will mix with the consciousnesses of others. Portions of your intent and purpose become theirs. (Roberts, *The Afterdeath Journal of an American Philosopher*, p. 10)

Every author you read, every contact you make, every conversation you have, mixes with your own consciousness to change it, in so far as you allow. In the same way, parts of your consciousness are used by others for their own needs and in their own fashion. Seth emphasizes that our conscious desires and intents attract components of consciousness from other identities, but these other components cannot invade or intrude upon our consciousness without our consent. Each identity possesses an integrity that cannot be violated without permission, and it closes its boundaries if the incoming information does not serve its purposes. Seth says there are no exceptions to this sanctity of the identity.

Summary

- ☐ In spite of the fact that consciousness is knocking on the door of physics, there still is no theory of consciousness.
- ☐ There is evidence that waking consciousness is but one aspect of our total consciousness, and that accessing other parts is possible by shifting to an alternative focus.
- ☐ The process of shifting the focus of consciousness does not extinguish it.
- ☐ Seth says that we are not our consciousness, but that it is an attribute used by each gestalt of consciousness units to perceive various dimensions of reality.
- ☐ Flashes of insight and intuition often enter into our conscious experience through the "backdoor" of the mind.
- ☐ While individual identity remains inviolate, each consciousness is connected to all consciousness, and the growth that is possible through those connections is limited only by the purposes and intent of the self.

Because there is neither theory nor measure of consciousness, these provocative ideas are largely ignored by scientists. However, the role of consciousness in creating reality seems inescapable when we confront some of the deepest questions raised by modern physics. Perhaps it is time for these challenging and fascinating concepts to be examined by the scientific community in a new light.

20 Final thoughts

Our considerations thus far indicate that the "something" out there is not an objective universe but a conglomeration of physical potentials. It is important to realize that these potentials are not predetermined to become physical events and are not equally probable. Some events have more propensity to stiffen, so to speak, into mass. Here the probabilities of quantum theory, along with Heisenberg's potentia, begin to take on meaning as a description of the realm of the wave function which contains all possibilities of events for a quantum system.

In his award-winning book *Taking The Quantum Leap,* Fred Alan Wolf describes the world of the wave function as analogous to an infinite table laden with every imaginable dish and arranged for immediate access to the food. It is as if All That Is is a colossal mother who generously invites all her children to eat in order to grow big and strong. But at this table, miraculously, every time one of the children reaches out and takes something, the very act of selection creates new platters of edibles. In like manner, all gestalts of consciousness not only enjoy partaking of the tremendous choices available, but also, in partaking, contribute to them. The counterpart in physics is that every time an electron makes a choice, a whole array of new probabilities is created.

This concept reverses the fundamental precept of classical physics (firmly fixed for the past several hundred years) that matter is basically inert, and which theorizes that if the right combination of particles come together, by some unspecified and amazing process, consciousness arises. In this reductionist model — that consciousness is created from particulate matter — the causal arrow is always from the bottom up, and the task of physicists is to disassemble matter into its most basic constituents. But in the process of discovering smaller and smaller hunks of matter, scientists found the most basic elements to be incompletely defined potentialities rather than particles. As we have seen, the "official" interpretation of this peculiar situation clearly states that an observer is necessary to bring these potentialities to actuality, and thus consciousness enters the picture.

□ ○ □

As far as the classical physicist is concerned, the past has been experienced, is complete, and cannot be undone. The future is also certain, in that physical laws predict mechanical interactions. Thus, in this view we live in a causal world where the past and future are determined. A given moment unfolds in the dimension of time, but events are established and are just there. However, in quantum theory, that viewpoint is no longer tenable. While the evolving wave function is classically determined, the actualization process — the quantum jump — is in the probabilistic realm. Almost everybody recognizes that quantum theory suggests an open future, but what about the past? Regarding his delayed choice experiment (pp. 144–145), Wheeler comments:

> It is wrong to think of that past as "already existing" in all detail. The "past" is theory. The past has no existence except as it is recorded in the present. By deciding what questions our quantum registering equipment shall put in the present we have an undeniable choice in what we have the right to say about the past.
>
> . . . The phenomena called into being by these decisions reach backward in time in their consequences back even to the earliest days of the universe. Registering equipment operating in the here and now has an undeniable part in bringing about that which appears to have happened. (Wheeler and Zurek, p. 194)

It would appear, then, that the past is as open as the future. William Irwin Thompson has a metaphor for this condition which is most apropos (*Pacific Shift*, p. 65). Suppose one is driving a car with the usual front windshield and rearview mirror. If the driver looks forward, the future is stretched out before him. If he looks in the rearview mirror, the past is stretched out behind him. But suppose he decides to make a turn. As soon as he chooses a different future, a glance at his rearview mirror tells him he has automatically changed his past. So, the descriptions we construct of our ancestors tell us more about ourselves than about them. In short, we have probable pasts as well as probable futures. Collectively, once we have pretty much agreed on a probable past, that becomes our official past. But, of course, it can always change.

Amazingly, though our wave aspect allows us to change in copious ways and make numerous connections, our identities remain intact. In a similar manner, two electrons can interact through the merging of their wave functions and thus create a greater reality for themselves. Yet each electron keeps its own identity though it is changed by its instantaneous contact with the second electron; as we know from the theory of phase entanglement, each has lodged a piece of its consciousness with the other. Both have grown and thereby created a more extensive identity or quantum system. This awe-inspiring and life-enhancing concept provides a description of our method of fulfilling our need to grow toward our own aims and objectives, while at the same time taking a role in the development of the whole human race. This same process, of course, applies to all animal life, all plant life, and all gestalts of matter. Also let us remember that the death of a particle can be seen only as a withdrawal of the particulate aspect. The wave function aspect continues to exist on other levels. The same may be said for the human body and its consciousness: though the body may cease to exist, consciousness remains.

□ ○ □

One significant insight we have attained is that the real cannot be separated from the imaginary. We use the idea of levels to simplify and to bring some clarity to complex concepts, but we must remember that all levels are interpenetrated. Reality is one whole, mixing and merging in waveform fashion. All energy, by means of its inherent creativity, is learning to use that fascinating tool we call consciousness. But through it all, the *you* of you remains and is never lost.

Darwin's theory of evolution, based on the concept of natural selection, states that various species survive according to how well they adapt to the changing environment. In Richard Dawkins's *The Selfish Gene*, a twist on this theory suggests that we are survival machines, blindly programmed to preserve genetic material, an idea which leaves no room for altruism or cooperation. The view promulgated here is quite the opposite. It is my belief that reality is a totally cooperative effort. In order to create our own body, let alone our entire environment, we require and receive the cooperation of endless gestalts of consciousness. All the particles in our body are fulfilling their own propensities and intents to grow and flower, and at the same time are helping us. In the same way, our growth as individuals is accomplished in the service of the entire human race. Such cooperation is illustrated by the inspiring words of the Jesuit paleontologist Teilhard de Chardin:

> Some day, after we have mastered the winds, the waves, the tides and gravity, we shall harness . . . the energies of love. Then, for the second time in the history of the world, man will have discovered fire. (Weber, p. 127)

It is easy enough to say that love is the glue that holds reality together, one might counter, but contradictory evidence abounds. How can we believe that the human species is growing in a loving way while wars still blanket the earth and individuals commit horrendous crimes? Seth's response is that just as a toddler who throws food on the floor is not evil but merely undeveloped, so many adults have not attained sufficient growth to handle their freedom. Love can be misdirected through mistaken free choice, resulting in what we often call evil. This is the way Seth describes it:

> You realize that a tiger, following its nature, is not evil. [But] looking at your own species you are often less kindly, less compassionate, less understanding. It is easy to condemn your own kind.
>
> It may be difficult for you to understand, but your species means well. You understand that the tiger exists in a certain environment, and reacts according to his nature. So does man. Even his atrocities are committed in a distorted attempt to reach what he considers good goals. He fails often to achieve the goals, or even to understand how his very methods prevent their attainment.
>
> He is indeed as blessed as the animals, however, and his failures are the results of his lack of understanding. (Roberts, *The Nature of the Psyche*, pp. 206-207)

The same observation applies to the entire human race. We are all here to learn. Some of us are still in preschool; some have moved to more advanced grades. What is important to remember is that the natural foundation of our universe rests on free choice and love. As we develop, as we learn, both individually and collectively, perhaps we shall choose to become kinder, more understanding, and more compassionate.

BIBLIOGRAPHY:

Bell, J. S., *Speakable and Unspeakable in Quantum Mechanics*. Cambridge University Press, 1987, New York.

Bohm, David and Peat, F. David, *Science, Order, and Creativity*. Bantam Books, New York, 1987.

Charon, Jean, ed., *The Real and the Imaginary*. Paragon House Publishers, New York, 1987.

Cole, K. C., *Sympathetic Vibrations: Reflections on Physics as a Way of Life*. William Morrow and Company, Inc., New York, 1985.

Quoted in Coveney, Peter and Highfield, Roger, *The Arrow of Time: A Voyage Through Science to Solve Time's Greatest Mystery*. Ballantine Books, New York, 1990.

Quoted in Davies, P. C. W. and Brown, J., eds., *Superstrings: A Theory of Everything?* Cambridge University Press, 1988.

Quoted in Davies, Paul and Gribbin, John, *The Matter Myth: Dramatic Discoveries that Challenge our Understanding of Physical Reality*. Simon & Schuster/Touchstone, New York, 1992.

Dyson, Freeman, "Theology and the Origins of Life," lecture and discussion at the Center for Theology and the Natural Sciences, Berkeley, Calif., November 1982.

Quoted in Enz, Charles P. and von Meyenn, Karl, eds., *Wolfgang Pauli: Writings on Physics and Philosophy*. Springer-Verlag, 1994, New York.

Factor, Donald, ed., *Unfolding Meaning: A Weekend of Dialogue with David Bohm*. Foundation House Publications, Gloucestershire, England, 1985.

Feynman, Richard, *The Character of Physical Law*. The M.I.T. Press, Cambridge, Massachusetts, 1967.

Feynman, Richard P., *QED: The Strange Theory of Light and Matter*. Princeton University Press, 1985.

Quoted in Foster, David, *The Philosophical Scientists*. Dorset Press, New York, 1991.

Gamow, George, *One Two Three . . . Infinity: Facts and Speculations of Science*. Bantam Books, New York, 1947.

Gribbin, John, *Schrödinger's Kittens and the Search for Reality: Solving the Quantum Mysteries*. Little, Brown and Company, Boston, 1995.

Hawking, Stephen, *A Brief History of Time: From the Big Bang to Black Holes*. Bantam Books, New York, 1988.

Heisenberg, Werner, *Physics and Philosophy: The Revolution in Modern Science*. Harper & Row, New York, 1962.

Herbert, Nick, *Elemental Mind: Human Consciousness and the New Physics*. Penguin Books, New York, 1993.

Herbert, Nick, *Quantum Reality: Beyond the New Physics*. Anchor Press/Doubleday, Garden City, New York, 1985.

Quoted in Hoffmann, Banesh, *The Strange Story of the Quantum*. Pelican Books, Harmondsworth, England, 1959.

Quoted in Holton, Gerald and Elkana, Yehuda, eds., *Albert Einstein: Historical and Cultural Perspectives*. Princeton University Press, 1982.

Quoted in Hooper, Judith and Teresi, Dick, *The Three-Pound Universe*. Macmillan Publishing Company, New York, 1986.

Quoted in Jones, Shirley A., ed., *The Mind of God & Other Musings: The Wisdom of Science*. New World Library, San Rafael, Calif., 1994.

Quoted in Laurikainen, K. V. and Montonen, C., eds., *Symposia on the Foundation of Modern Physics 1982: The Copenhagen Interpretation and Wolfgang Pauli*. World Scientific Publishing Company, Singapore, 1993.

Quoted in Moore, Walter, *Schrödinger: Life and Thought*. Cambridge University Press, 1989.

Pagels, Heinz R., *The Cosmic Code: Quantum Physics as the Language of Nature*. Bantam Books, New York, 1983.

Penrose, Roger, *The Emperor's New Mind: Concerning Computers, Minds, and the Laws of Physics*. Oxford University Press, 1989.

Rae, Alastair, *Quantum Physics: Illusion or Reality?* Cambridge University Press, 1986.

Roberts, Jane, *The Afterdeath Journal of an American Philosopher: The World View of William James*. Prentice Hall Press, New York, 1978.

Roberts, Jane, *The Nature of Personal Reality: A Seth Book*. Prentice Hall Press, New York, 1974.

Roberts, Jane, *The Nature of the Psyche: Its Human Expression*. Bantam Books, New York, 1979.

Roberts, Jane, *Seth: Dreams and Projection of Consciousness*. Stillpoint Publishing, Walpole, N. H., 1987.

Roberts, Jane, *The Seth Material*. Prentice Hall, Inc., New York, 1970.

Roberts, Jane, *Seth Speaks: The Eternal Validity of the Soul*. Prentice Hall Press, New York, 1987.

Quoted in Rucker, Rudy, *The Fourth Dimension: Toward a Geometry of Higher Learning*. Houghton Mifflin Company, Boston, 1984.

Sachs, Mendel, *Relativity in Our Time: From Physics to Human Relations*. Taylor & Francis, Inc., Bristol, Pa., 1993.

Quoted in Weaver, Jefferson Hane, *The World of Physics*, Vols. II and III. Simon & Schuster, New York, 1987.

Weber, Renée, *Dialogues with Scientists and Sages: The Search for Unity*. Routledge & Kegan Paul, London, 1986.

Weinberg, Steven, *Dreams of a Final Theory: The Search for the Fundamental Laws of Nature*. Pantheon Books, New York, 1992.

Quoted in Weintraub, Pamela, ed., *The Omni Interviews*. Omni Press, New York, 1984.

Wheeler, John Archibald, *At Home in the Universe*. AIP Press, Woodbury, N. Y., 1994.

Wheeler, John Archibald and Zurek, Wojciech Hubert, eds., *Quantum Theory and Measurement*. Princeton University Press, 1983.

Wilber, Ken, *The Spectrum of Consciousness*. Theosophical Publishing House, Wheaton, Ill., 1977.

Wolf, Fred Alan, *Parallel Universes: The Search for Other Worlds*. Simon & Schuster, 1988, p. 126.

Quoted in Zajonc, Arthur, *Catching the Light: The Entwined History of Light and Mind*. Bantam Books, New York, 1993.

Quoted in Zukav, Gary, *The Dancing Wu Li Masters: An Overview of the New Physics*. William Morrow and Company, Inc., New York, 1979.

GLOSSARY:

algorithm A calculational procedure containing a well-defined sequence of operations.

All That Is A term from the Seth material meaning the infinite psychological realm whose energy is within and behind all formations. It is alive within the least of Itself. All of Its creations are endowed with Its own abilities, and it is infinitely becoming, never complete.

altered states of consciousness Beneath our ordinary mental life, a range of consciousness which provides alternative structures of reality. The most obvious is dreaming. However, other states may occur through psychedelic drugs, meditative techniques, sensory deprivation, trance, hypnosis, mystical experience, etc. Experiencing these states allows a subjective exploration of consciousness.

antiparticle The counterpart of an elementary particle with identical mass and spin but opposite electric charge and magnetic properties. When a particle and its antiparticle collide, they are destroyed, with the equivalent energy surviving as radiation. Some electrically neutral particles are their own antiparticles. Fortunately, antiparticles are usually seen only in laboratories; the rare exception is in the debris of a cosmic ray shower.

archetype A predisposition toward certain patterns of psychological performance that are passed down from our ancestral past (human and animal) and are linked to instinct. In Platonic terms, an archetype is the underlying form or idea of a thing. According to Jung, archetypes are forms without content in the collective unconscious, which become activated as psychic force.

Bell's theorem (or inequality) The correlations discovered by John Stewart Bell between simultaneous measurements of two widely separated particles, in which a limit is observed if reality is local (governed by local interactions). When experiment found the limit to be violated, it was generally agreed that any model of reality must be nonlocal. *See* locality/nonlocality.

big bang theory The theory that the universe was created by a gigantic explosion from a singularity (*see below*) about eighteen billion years ago. Its main confirmation is the detection of black-body background radiation left over from the original event. The theory states that no space or time existed before the original bang.

black body An object capable of efficiently absorbing all frequencies of electromagnetic radiation and emitting all frequencies when brought to incandescence. A true black body is a perfect absorber and radiator. Max Planck discovered the quantum while studying theoretical problems connected with black-body radiation.

black hole A material body that has collapsed to a highly compressed state in which the gravitational field is so intense that not even light can escape from its surface. It is characterized by only three properties: mass, charge, and angular momentum (although Stephen Hawking has shown that quantum effects do permit some radiation). **white hole** A hypothetical time-reversed black hole. While a black hole absorbs all matter, a white hole is a source for all matter. Mathematically it is possible that a black hole is connected to a white hole through a tunnel that may end up in another part of the universe or in a second universe.

causality Interrelationship between cause and effect. If an effect is a necessary result of a given cause, then causality is operating. However, causality must account for *all* factors governing the change. Since the view presented here is that all systems (including the universe) are abstractions from a greater whole, then all factors are not known, and causality must be conditional.

collective unconscious Jungian term for universal, shared aspect of the unconscious mind, which contains patterns of instinctual behavior (archetypes). The collective unconscious is manifest in the personal unconscious in the form of images and instincts.

complementarity principle The statement by Niels Bohr (and a major element in the Copenhagen interpretation of quantum theory) that complementary properties of the same object exist such that if one property is known, knowledge of the other property is excluded. Therefore, an object can be described in two mutually exclusive ways and yet not have a logical contradiction. The best example is that an electron can be described as both particle and wave.

consciousness unit (CU) Seth's term for the smallest units of consciousness — "unit" being a term of convenience, since, according to Seth, energy is not divided. CUs have their source outside space-time and are not affected by our limits on the velocity of light; they slow down to our velocity of light to create physical structure. CUs are the source for *all* consciousness and are endowed with unpredictability, which allows for infinite patterns and fulfillments.

coordinate points In Seth's concept, points where other realities coincide with our own and act as channels through which energy flows; invisible paths from one reality to another. In a manner similar to that in which electrical transformers change the volt-ampere relationship, coordinate points transform thoughts and emotions into matter. There are three kinds of coordinate points: absolute, main, and subordinate.

dark matter Matter hidden throughout the universe that emits no radiation for instruments to detect. Its existence is inferred from astronomical measurements that fall far short of the amount of matter needed for the required gravitational attraction between and within galaxies.

determinism The philosophical doctrine that assumes that if the state of a system is known at a given time, and if the physical laws governing its behavior are known, all future states can be determined, at least in principle. Most physical theories prior to quantum theory have been deterministic in this sense.

EE units Units of electromagnetic energy, which, according to Seth, emanate from all gestalts of consciousness. As the electromagnetic expression of subjective experience, EE units are emitted by consciousness and propelled into physical actualization if the intensity of the feeling, thought, or emotion from which they arise is sufficient. If not, they can be seen as latent matter. EE units exist at many velocities, but they are perceived by us only at our velocity of light.

entropy/information Entropy is the measure of energy not available to do work; the higher the entropy, the more disordered the system. Information generates new orderly structures, thus decreasing entropy. Therefore, entropy and information are complementary.

ether A medium hypothesized by classical physicists to fill space and carry electromagnetic radiation. However, when its properties were calculated, the ether had to have both the qualities of a tenuous gas and a rigidity greater than steel. This nonviable concept was replaced with Einstein's concept of space-time.

events Points in four-dimensional space-time, which do not exhibit extension or duration. In the Newtonian view, the universe consists of things, whereas in the Einsteinian view, the universe consists of events, e.g., the interaction of subatomic particles.

explicate order David Bohm's term for the domain referred to by Cartesian coordinates (used to designate location in space-time). It displays the separateness and independence of fundamental constituents and is manifest or visible (directly or with instruments). It is secondary to the implicate order (*see below*), which unfolds to create the explicate order.

fact-space Rudy Rucker's name for a space that arranges our thoughts and sensations in multiple dimensions, each of which is a possible category or distinction. That is, for any question we might pose about an object, there is a range of possible answers, and each range is a dimension in fact-space. Examples of ranges are hot/cold, sweet/sour, etc. Rucker comes to the conclusion that fact-space and Hilbert space (*see below*) are similar.

field An area of space characterized by a physical property that is normally invisible and intangible but under certain circumstances can interact with matter. Classically, the action between two material objects separated in space is described in the language of fields. In quantum field theory, the interaction is viewed as an exchange of so-called messenger particles, which conveys a particular force. Examples of such particles are the photon (electromagnetic) and graviton (gravitational).

Fourier analysis A mathematical technique for analyzing a periodic function (e.g., a musical note) into its fundamental and harmonic components.

frame of reference A set of coordinate axes by which the position of an object can be specified.

frameworks Seth's term for interpenetrating levels of existence. Our outer reality is Framework 1, and our inner reality and the creative source from which we form events is Framework 2. We draw from Framework 2 by focusing our attention properly. All That Is contains an infinite number of frameworks.

geons A particular arrangement of a sufficiently great concentration of energy (electromagnetic or gravitational) which moves through space as a unit and exerts a gravitational attraction like ordinary mass, although it is only empty, curved space. This concept of building mass out of pure geometry was originated by John Wheeler.

Hilbert space The abstract space used by physicists to describe quantum mechanical states. A single point in Hilbert space represents the entire quantum system. The wave function is represented by a vector (state) composed of a superposition of states, each corresponding to a possible result of a measurement. A vector is a quantity that has direction as well as magnitude.

hologram A photographic record created by laser light directly reflected from an object and laser light going directly to the plate to form a complex interference pattern. When the plate is illuminated with coherent light, the wave front produced creates a three-dimensional image of the original object. In this way, the whole image can be recreated from each portion of the record.

implicate order The basic order, according to David Bohm, from which our three-dimensional world springs. It is multidimensional, and its connections are independent of space and time. The implicate order is identified with the wave function in quantum theory. *See* explicate order.

light As used in this book, the entire electromagnetic spectrum. Light can be interpreted as a wave, as in classical theory, or as a stream of quanta or photons, as in quantum theory. The frequency of the wave is related to the energy of the photon by Planck's constant. The velocity of light in free space is constant regardless of the motion of the observer.

David Bohm sees the implicate order as an ocean of energy, or light. Matter is a condensation of this energy; it arises when light rays reflect back and forth to form a pattern. Without reflection, there is simply pure light. Matter carries with it time and space, but pure light does not.

Seth states that the electromagnetic spectrum that we observe is a small portion of an infinite spectrum. The attribute that separates sections of the spectrum is the velocity of light. For light to become manifest as physical form, its velocity must be increased or decreased to our range (300,000 km/sec).

linearity/nonlinearity The relationship between input and output in a system. In linear systems the input is always directly proportional to the output; linear systems can be separated into parts so that the whole is exactly equal to the sum of the parts. In nonlinear systems, a given input can cause unexpected outputs, and the whole can be greater than its parts.

locality/nonlocality The condition that defines the causal relationship between events. In a local reality, information cannot travel faster than light. In a nonlocal reality, objects can influence each other instantaneously. Bell's theorem (*see above*) indicates that if the world is made up of separate objects, they must have nonlocal connections.

mechanistic philosophy The idea that the universe is composed of basic entities that follow an absolute set of quantitative laws such that if these laws were known and the intellect were great enough, all future events could be calculated and predicted with complete precision. This idea emerged from an extrapolation of Newtonian mechanics to other domains of knowledge.

moment point Seth's term for the present moment, the point of interaction between all existences and realities, through which all probabilities flow. According to Seth, the past, present, and future exist together and can be experienced in a moment. This is analogous to Minkowski's space-time in which all points of space and all points of time exist en bloc. The moving present we perceive as time is, in Seth's concept, a projection from a higher-dimensional timeless order.

phase space A space with a large number of dimensions. A single particle in a physical system requires three position coordinates and three momentum coordinates, or six in all, so a system with n particles requires a phase space of $6n$ dimensions. Thus, a single point in phase space can represent an entire state with any number of particles.

Planck's constant A fundamental constant in quantum theory, designated by the letter h, defining the relationship between the energy carried by a photon and its frequency. Planck's constant requires an observable measurement to take on specific values rather than any range of values.

plenum Space that is completely filled with matter/energy; the opposite of vacuum.

quantum The smallest discrete unit of energy released or absorbed in a process. The quantum for electromagnetic energy is the photon.

quantumstuff A term used by Nick Herbert to suggest that the world is ultimately composed of one substance. The classical physicist divided the world into two entities, matter and fields. Quantum theory removes the distinction between the two by combining the particle and wave, i.e., quantumstuff.

quantum theory A mathematical theory that describes the behavior of all systems (in principle) but is especially useful in the atomic and subatomic realms. According to quantum theory, *individual* events are inherently unpredictable, nonseparable, and in the view of some physicists, observer-dependent.

quantum wave *See* **wave.**

reductionism The principle that seeks to explain complex phenomena in terms of simpler ones. In the reductionist view, psychology can be explained by biology, which is reduced to chemistry and ultimately to particle physics. Reductionism denies the possibility of a collective property that supercedes and cannot be explained by the component parts.

relativity theory, general The theory that explains the difference between accelerated and nonaccelerated systems and its relationship to gravity. As a consequence, space-time is seen not as rigid but as elastic, with its curvature determined by matter and energy, and gravity becomes a product of geometry rather than a force. These aspects of relativity theory resulted in the concepts of black holes, finite but unbounded universes, and ruptures in space-time. **special** The theory that postulates that the velocity of light is constant for all observers or sources. Also, physical laws are the same in all inertial frames of reference. From these assumptions, Einstein deduced the equivalence of mass and energy and the elasticity of space and time.

renormalization A rescaling procedure used to subtract infinities so that finite answers are obtained. In attempting to unite quantum and relativity theories, physicists' calculations of the interaction of photons and electrons produced infinite answers, which are not acceptable. This problem is controlled by the renormalization procedure.

rest mass The measured mass of an object when the object and the observer are both at rest in the same frame of reference. According to the special theory of relativity, a body in motion relative to an observer at rest increases its mass, especially at very high velocities. The mass that the observer measures in this case is called the relativistic mass.

Schrödinger's equation An equation that describes the time evolution of a quantum state as represented by a quantum wave function (*see below*). It is a complex differential equation which is completely deterministic, but when a measurement is made, the quantum state is abruptly changed to one of a set of new possible states for which the probability of occurrence can be computed from the wave function.

singularity A boundary to the three-dimensional universe where the laws of physics may not be operable; e.g., the center of a black hole, which is infinitely dense and thereby has ripped a hole in space-time.

soma-significance A term used by David Bohm to more clearly explain the age-old problem of the relationship between mind and body. In his view, soma represents the physical aspect of a body while significance is the mental aspect. However, the two are really aspects of one overall reality. Each reflects and implies the other.

space-time A four-dimensional mathematical construction, based on the work of Minkowski and Einstein, which describes our world more accurately than the three-dimensional concept. This construction arose naturally from relativity's length contraction and time dilations at high velocities. What we call space *and* time can be understood as projections from the space-time world, and they vary according to the velocity of the object.

space-time interval A combination of space and time intervals that are invariant for all observers. Space intervals and time intervals are not invariant when separated. The invariance of the space-time interval indicates that space and time are fused together.

uncertainty principle The statement by Werner Heisenberg that it is impossible to measure a pair of noncompatible observables (e.g., position and momentum) both precisely and simultaneously because all energy comes in quanta and all matter has both wave and particle aspects. An accurate measurement of position requires a short wavelength (high energy), and an accurate momentum measurement requires a long wavelength (low energy).

virtual particle A quantum particle that appears and disappears spontaneously and whose existence is of extremely short duration. Heisenberg's uncertainty principle allows for this phenomenon, and as a result, these short-lived particles are considered to be different from the more familiar observable particles.

wave A disturbance moving through a medium. The resulting displacements which constitute the wave are determined by the elastic and inertial factors of the medium. However, the **quantum wave** travels in a sea of probabilities and potentials which does not have measureable energy and is outside of space-time. Therefore, its amplitude is related to a probability rather than the disturbance of a material substance.

wave function A mathematical description of the state of a quantum system; a solution of Schrödinger's wave equation.

wave-particle duality The twofold nature of the quantum objects, which can display wavelike and particlelike properties.

white hole *See under* **black hole**.

zero-point energy The irreducible quantity of energy, even at the minimum energy possible, of the discrete states of the quantum field; a result of the uncertainty principle.

INDEX:

A

B

B—cont.

C

D

E

F

G

H

I

I—cont.

J

K

L

M

M—cont.

N

O

P

P—cont.

Q

R

R—cont.

S

S—cont.

T

U

U—cont.

V

W

W—cont.

Z

Seth Network International (SNI) is a network of Seth/Jane Roberts readers from over 30 countries who meet to explore ideas from the Seth material. For further information, please contact:

Seth Network International
P.O. Box 1620
Eugene, OR 97440 USA
Phone (541) 683-0803 Fax (541) 683-1084
http://www.efn.org/-sethweb